Automatic Chemical Analysis

Automatic Chemical Analysis

PETER B. STOCKWELL

P. S. Analytical Ltd, St Paul's Cray, Orpington, Kent

in association with **WARREN T. CORNS**

UK Taylor & Francis Ltd, 1 Gunpowder Square, London EC4A 3DE
USA Taylor & Francis Inc., 1900 Frost Road, Suite 101, Bristol, PA 19007

Copyright © Taylor & Francis Ltd 1996
All rights reserved. No part of this publication may be reproduced stored in a retrieval system, or transmitted, in any form or by any means, electronic, electrostatic, magnetic tape, mechanical, photocopying, recording or otherwise, without the prior permission of the copyright owner.

British Library Cataloguing in Publication Data
A catalogue record for this book is available from the British Library
ISBN 0-7484-0480-5 (cased)

Library of Congress Cataloging Publication Data are available

Cover design by Amanda Barragry

Preliminary material typeset by Solidus (Bristol) Ltd

Printed in Great Britain by T. J. Press (Padstow) Ltd

Contents

Preface		1
1	**Introduction to automatic chemistry**	5
	1.1 Why automate?	6
	1.2 Examples of successful automation	10
	1.3 Effects of automatic analysis	16
	1.4 The analytical chemist	16
	1.5 Laboratory management	17
	References	17
2	**Principles of automatic analysis**	19
	2.1 Introduction	19
	2.2 Education, specification and communication	22
	2.3 Developments in the philosophy of automation	27
	2.4 Principles of automatic instruments	28
	2.5 Automatic continuous analysers	47
	2.6 Conclusions	62
	References	64
3	**Total systems approach to laboratory automation**	67
	3.1 Case Study I - Regulation of tar and nicotine in cigarettes	67
	3.2 Case Study II - Mercury analysis in a natural gas plant	86
	3.3 Case Study III - Water laboratory automation	95
	References	100
4	**Automatic sample preparation and separation techniques**	101
	4.1 Introduction	101
	4.2 Liquid-liquid extraction (solvent extraction)	101
	4.3 Flash vaporization in continuous-flow analysers	104
	4.4 Chromatographic applications	107
	4.5 Automation of sample preparation of foodstuffs for trace metal analysis	120
	4.6 Recent developments in automatic sample preparation	126
	4.7 Alternative approaches	132
	References	133

Contents

5	**Automated atomic spectroscopy**	137
	5.1 Introduction	137
	5.2 Hydride/vapour-generation techniques	139
	5.3 Flow-injection and preconcentration applications	145
	5.4 On-line dilution systems	153
	5.5 Sample introduction into ICP-AES or ICP-MS by electrothermal vaporization	155
	References	159
6	**Robots and computers**	163
	6.1 Introduction	163
	6.2 Robot types and functions	164
	6.3 Advantages and justification of robotics in the laboratory	167
	6.4 Applications of robotics	170
	6.5 Robotic automation of total nitrogen determination by chemiluminescence	180
	6.6 A validated robotic system for dissolution assay of an immediate-release/sustained-release tablet	182
	6.7 Automated extraction and analysis of chlorinated compounds by large volume on-column injection GC/MSD	183
	6.8 Collection and determination of doses delivered through the valves of metered aerosol products	186
	6.9 A solid reagent handling robot	188
	6.10 Conclusions	191
	References	191
	Appendix: Robotic companies with products listed in Table 6.1	194
7	**Examples of automatic systems**	195
	7.1 Automatic specific gravity and refractive index measurement	195
	7.2 Computer-assisted analysis of metals	200
	7.3 Automated tablet dissolution systems	204
	7.4 Routine determination of mercury at low levels	207
	7.5 Automated pH and conductivity measurement	217
	7.6 Summary	219
	References	220
8	**The future of automatic analysis**	221
	8.1 Introduction	221
	8.2 Influence of computer power	221
	8.3 Systems validation and credibility	226
	8.4 Conclusions	228
	References	228
Glossary of manufacturers' addresses		229
Index		233

Preface

The first edition of *Automatic Chemical Analysis* was published in 1974 and was written in collaboration with the late Jim Foreman. It was well received and became a set course book in many universities. Its aim was to provide a basic grounding in the subject and to cover the wider issues of economic assessments, educational requirements and specification of systems. There are now other books on laboratory automation in print [1,2], but none of these will help potential users of automation to evaluate the economic viability of a project; indeed, they may even put them off. The other books are scholarly rather than practical.

The first edition of *Automatic Chemical Analysis* is out of date and clearly the time has come for a new edition. The following is a brief description of my career, which I hope will help readers to understand my approach to automation. The advice offered in this new edition is based on proven practical applications which have been in use in the real world for many years.

I have been extremely lucky in my career in that I have been exposed to many areas of laboratory life: in academia, in a government laboratory and in industry, both as a part of large organizations, and, more recently, with my own design and development company.

Throughout the years, my experience in the use, design and development of instrumental and automated techniques has been extremely wide.

I began my doctorate at Imperial College expecting to be engaged purely in physical chemistry; however, I was surprised to find myself involved with analytical chemistry. Over the three years of research, I developed my own gas chromatograph, vacuum and computer programs and I was introduced to the fields of infra-red spectroscopy, nuclear magnetic resonance and mass spectroscopy [3]. I was then offered the opportunity to join Philips Research Laboratories to assist with the introduction of a new range of gas chromatographs. The group I joined at Philips, which was essentially set up to research gas chromatography, had the good fortune to be located within the engineering workshop area. One of my colleagues was able to get jobs done in the engineering department, and the enthusiasm he created amongst these staff meant that our projects progressed rapidly. The organization's major strengths were in electronic design and theoretical aspects and the small group working on chromatography was rather out of place. However, the skills of the electronics specialists were extremely valuable when, for example, the use of Curie point pyrolysis was explored as an introductory system for polymers. Simon and co-workers [4-7] presented data on the use of Curie point pyrolysis to couple with a gas chromatograph using an eddy current heater with an output of 3 kW. A study of the problem, and the requirements of the eddy current heater to rapidly heat a wire to a fixed temperature, showed that the 3 kW heater was over-powered and required the addition of a sophisticated timer circuit to drop the power supplied. With the help of the electronics group and the mechanical design team, I was quickly able to design and develop a simple

50 W Curie point heater system to form a sample introduction system for solid samples into a gas chromatograph. The product won a design award in the USA and the device is still selling some 25 years later - this is because it is easy to use, sensibly priced and it works. The design and other work tackled in this job was educational, both because of the experience I gained in electronics, mechanical engineering and design, and also because of what I learned about the choice of materials for design problems.

The real world of industry was brought home to me during a problematic merger with Pye Unicam of Cambridge, and I left Philips in 1967 to join the Laboratory of the Government Chemist (LGC). It was a good time to join the LGC. I was beginning to be interested in automation and a Technicon AutoAnalyzer System, originally designed for clinical chemistry, had just been purchased. Working with Ron Sawyer, I gained my first experience of automation with real analytical applications. The LGC wanted to automate and had the routine analytical requirements to justify an investment in automation. My experience in instrumentation was useful here, and not only could we provide an automated solution to a problem, but we were also able to engineer the equipment. Our products had to work not only when operated by the designer, but also when used by semi-skilled staff.

In 1969, Ron Sawyer moved to become Director of the Customs and Excise Division, and the management were sufficiently enthusiastic to set up a group devoted to laboratory automation. The economics of automation were important and between 1969 and 1980 a great number of analytical problems were cost-effectively solved.

In 1976, I took a sabbatical year to teach analytical chemistry at Purdue University in Indiana. Professor Harry Pardue was kind enough to sponsor me there, and the period gave me an insight into new techniques, such as diode-array spectrometers and microcomputers. I was also able to make many friends in industry and academia and to gain an insight into the major differences in style, approach, and training, between the USA and Europe. At this time I was also able, with a great deal of backing from many friends and colleagues, primarily Professor Howard Malmstadt (then at University Urbana Champaign, Illinois, USA), Dr Rolf Arndt (then at Mettler-Toledo AG, Switzerland), and Dr Fred Mitchell (then at the Clinical Research Centre, UK) to start the *Journal of Automatic Chemistry*.

In 1978, I became Director of Research at the LGC. In addition to responsibilities for automation and computerization, my division was responsible for analytical instrumentation, for example NMR and inductively coupled plasmas. As part of this job, I collaborated with the UK's Department of Industry sponsorship scheme promoting the development of analytical instrumentation. Thus, I was actively involved in vetting proposals, modifying them and then controlling them to ensure a good return on investment. At this time, the introduction and availability of microprocessors had become an extremely taxing and topical area of interest. During my years at the LGC, I tackled all aspects of automation, from simple mechanical devices [9] to fully hierarchical computer applications [10].

In 1980 LGC, I moved into industry as Technical Director of Plasma Therm Europe. My main expertise was and continues to be, in analytical instrumentation, whereas the main business of Plasma Therm was in the emerging area of semiconductor process instrumentation. The company had an interest in analytical chemistry through its ability to design RF plasma generators. These form an ideal source for inductively coupled plasmas, which provide a valuable analytical spectroscopic tool. Setting up an R & D and production facility in the UK, whilst assisting in organizing a distributor network in Europe, was rewarding and exciting. However, my main interest and skills were in analytical chemistry and in 1983 it became obvious that these avenues offered more personal satisfaction than did semiconductors.

In 1983, I set up my own company: P.S. Analytical. A broad range of analytical modules with hardware and software have been designed, developed and manufactured by the company and are sold internationally. The main expertise of P.S. Analytical is to research a problem and to provide, at an economic cost, a fully engineered solution for routine use. The company collaborates with other companies, with universities and with government laboratories. P.S. Analytical has exposed me to different areas of analytical chemistry and has allowed me to tackle many areas of analytical application, for example sample types. I have been actively involved in all aspects of the automation, especially economics, philosophy and management. P.S. Analytical has flourished and increased its range of products from simple accessories to fully integrated instrumental systems, for example, the Merlin Plus Fluorescence system for the determination of ultra-trace levels of mercury.

This book has been written primarily for practising analytical chemists working in research and in industrial and clinical environments, but will also be an invaluable aid to students and teachers.

Acknowledgements

It is a pleasure to acknowledge those people who have contributed to the production of this book. Without the enthusiasm of Ellis Horwood whose company published *Automatic Chemical Analysis* in 1974, I do not think that this revision would have even been started. I am also indebted to Michaela Lavender who has worked with me for several years, guiding the style of this book and with the *Journal of Automatic Chemistry*. At the outset of this project Dr Warren Corns provided much of the literature survey work and his input towards Chapter 5 has been extremely valuable. I have been encouraged and helped by many friends and relatives into completing this book, their support cannot be easily repaid. Matthew Pritchard provided many of the illustrations used. I would also like to thank my daughter-in-law Avril, who typed the initial drafts, co-ordinated the proofs and in addition provided camera-ready copy for the publishers. Her patience with the many changes is gratefully acknowledged. Finally I wish to thank the publishers, Taylor & Francis, for taking on this project with enthusiasm and a spirit of co-operation.

REFERENCES

[1] Valcarcel, M. and Luque de Castro, M.D., *Automated Methods of Analysis*, Elsevier, The Netherlands, 1988.
[2] Cerda, V. and Revmis, G., *An Introduction to Laboratory Automation*, Wiley, Chichester, UK, 1990.
[3] Stockwell, P.B., Ph.D Thesis: Kinetic Studies on Ditrifluoromethyl-hexafluorocyclobutane. University of London, UK, 1965.
[4] Simon, W. and Giacobbo, H., *Angewandte Chemie*, 1965, 11, 938.
[5] Simon, W. and Giacobbo, H., *Chemie-Ingenieur-Technik*, 1965, 37, 709.
[6] Giacobbo, H. and Simon, W., *Pharmaceutica Acta Helvetiae*, 1964, 39, 162.
[7] Vollmin, J., Kreimeer, P., Omura, I., Seibl, J. and Simon, W., *Microchemical Journal*, 1966, 11, 73.

References

[8] *Journal of Automatic Chemistry*, Published by Taylor & Francis, 1 Gunpowder Square, London EC4A 3DE, UK.
[9] Sawyer, R., Stockwell, P.B. and Tucker, K.B.E., *Analyst*, 1970, 95, 284.
[10] Stockwell, P.B., *Journal of Automatic Chemistry*, 1979, 1, 216.

1

Introduction to automatic chemistry

Automatic analysis is well established as a subject area, but despite the decade or more since the publication of *Topics in Automatic Chemical Analysis*, there is still no adequate definition of the area the subject embraces. This is not surprising when one considers the length of time and number of in-depth discussions that various groups, such as the International Federation of Clinical Chemistry (IFCC), have to spend on defining evaluation protocols and the like [1]. The International Union of Pure and Applied Chemistry (IUPAC) has also sought, through its Commission on Analytical Nomenclature, to offer rigorous definitions and to provide a common international terminology. IUPAC define automation as 'the use of combinations of mechanical and instrumental devices to replace, refine, extend or supplement human effort and facilities in the performance of a given process, in which at least one major operation is controlled without human intervention, by a feedback mechanism'. Mechanization, on the other hand, is defined as 'the use of mechanical devices to replace, refine, extend or supplement human effort'. The distinction between the two terms is quite clear according to IUPAC, insofar as 'automation' describes systems that involve a feedback loop. Whilst this is quite logical, it is not used in this book because a feedback unit is not vital to an analytical chemist. The major thrust for research and development in automation and/or mechanization was largely twofold:

1. To improve the cost-effectiveness in discharging large analytical workloads, especially those of a repetitive nature.
2. To improve the method performance, notably precision, in these circumstances. The advances made have been commonly termed 'automation'.

Microprocessors, computers and robots are in themselves only subsets of the general field of automation. When the *Journal of Automatic Chemistry* was launched in 1979, one of my fellow founders, Professor Howard Malmstadt, suggested that it would be many years before the journal would be understood in the USA. This, he believed, was because the term 'computing' was generally considered synonymous with 'automation'. The reality that sample preparation was often the major bottleneck in analysis was ignored. More recently, robotics have found considerable favour, and, again, the term 'automation' has been confused to refer to robotics and nothing else.

The principal operations involved in a chemical analysis are set out in Fig. 1.1. In considering the benefits of automating an analysis, each step in Fig. 1.1 must be taken into account. Sampling requirements will often not be amenable to automation and the task of reporting the result to the end-user may not be clearly defined. Analysts are often not involved in either the choice of sample nor in the reporting of the result - I believe that it is important that they should be. For example, a sample which is not representative of the material it comes from will be of no real value, just as the reporting of a result without full justification of its precision is pointless.

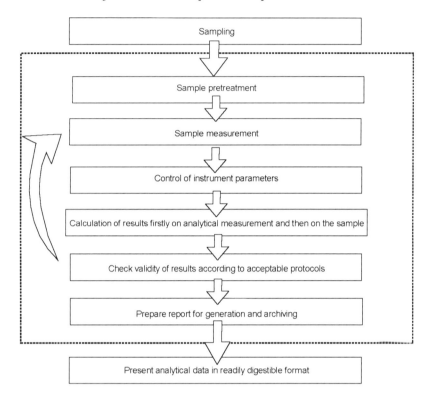

Fig. 1.1   The steps involved in analysis. The steps enclosed within the dashed lines are normally under the control of the analyst. Those outside the dashed lines are not.

Very few publications in the field of automatic analysis have embraced all the steps outlined in Fig. 1.1. The sample measurement stage has received the most attention, and almost all measurement techniques used in chemical analysis have been partly or fully automated. Indeed, automating this stage in respect, for example, of Fourier Transform infra-red or the application of near infra-red spectroscopy have only become viable through the introduction of inexpensive, but powerful, computing facilities. More recently, attention has been focused on the post-analysis stages: data processing and reporting using computers and data-presentation software packages have been extensively studied. It should be remembered that reports generated with bar or pie charts often present numerical data in a far more easily digested form for the customer and user of the data.

1.1   WHY AUTOMATE?

The reasons for installing automatic analysers to replace manual methods are many and varied, and their priority will be related to the nature of the organization of which the laboratory is part. This is discussed in more detail later in this chapter, together with examples. The principal motivation for most organizations is economics; the effect of automating can be assessed both as an internal laboratory study, or for the company as a whole, particularly in the industrial environment. In the latter case, automatic analysis can improve efficiency by reducing or eliminating the need to store intermediates or the final product, and the financial advantage gained is likely to far exceed the cost of the automatic equipment. When the economic justification for automation must be made internally, which would be the case for a consultant analytical laboratory, the true cost of manual and automatic analysis must be calculated, to determine whether the additional output, or staff savings, from automatic analysis outweigh the cost of the automatic equipment over a reasonable period. It is probably worth noting that the savings incurred are not likely to result in staff losing their jobs, and any fears of this nature must be quickly dealt with by the management. Automation simply releases staff from mundane chores, so that they can use their skills and training on more meaningful tasks which cannot be undertaken by automatic analysers. There have been few publications dealing effectively with the economics of analysis.

1.1.1 Cost benefit analysis

The costs of a manual analysis can be approximated as shown in Eq. 1.1.

Eq. 1.1   Cost of analysis - manual procedure

Cost = $F_m (A_1T_1 + A_2T_2 + A_3T_3 + A_4T_4 + A_5T_5...)$
F_m = Manual tariff factor - converts to a monetary unit
A_1 = Activity, i.e. staff involvement
A_2 = 1 when staff are completely involved
A_3 = Reduced significantly when staff are simply loading a sample onto a tray or drying a sample in an oven overnight

T_i = Time involved in activity

Total cost can be reduced by reducing either Time or Activity

---

The A factors for each stage represent the operator's time - for drying and incubation it is very small, but it can be high in procedures such as weighing. F is a tariff factor, incorporating staff and overhead costs, to convert the activity and time factors into the real cost in cash terms. From Eq. 1.1 it is clear that there are two ways of reducing the cost of an analysis. Firstly, reducing the analysis time - this is often put forward as a prime justification for automation. The reduction in time also leads to an increase in analytical capacity, and, in some cases, this capacity can be of value in providing additional revenue. Alternatively, the costs can also be reduced by lowering the A factor, even at the expense of increasing the time required for the task. For many years at the Laboratory of the Government Chemist (LGC), attention was drawn to the manner of analysis of the moisture in tobacco for duty assessment. Nuclear Magnetic Resonance (NMR) and other techniques were considered as replacements for the statutory procedure. In an economic assessment of the two approaches, the statutory process had minimal costs. It involved simply weighing out a fixed weight of sample and placing it in a muffle furnace overnight. On return the next morning, the samples were reweighed and the loss of weight, and, hence moisture level, was obtained. In this process the A

factor was minimal and replacing the process with an automatic procedure could not be justified. A similar situation arises in the determination of total mercury in aqueous samples as described by Stockwell and Godden [2]. The digestion step outlined in the Yorkshire Water *Methods of Analysis Handbook* [3], to convert any organically bound mercury, can be carried out in a continuous flow manifold activity. Leaving samples overnight to digest will prepare them for analysis the next morning by a simpler system - which is less expensive to develop and much more reliable.

For the same analysis carried out fully automatically the cost is given by Eq. 1.2.

Eq. 1.2   Cost of analysis - fully automatic procedure

Cost = $F_A(A_R T_R + A_S T_S)$
F_A = Automatic tariff factor converts to a monetary value
A = Staff involvement in activity
T = Time involved in activity
R = Running automatic system
S = Service

If the method is not fully automatic, any stage which remains manual can be costed in Eq. 1.1 and carried forward into Eq. 1.2. It is important to include the cost of servicing to keep the system running smoothly and reliably. Of course, the design of an automatic instrument should take this factor into account. Too complicated a design will often result in an increased maintenance and will therefore reduce the overall merits of automating. Many years ago, when the original patent protecting the Technicon AutoAnalyzer had expired, several multi-channel instruments based on a batch procedure were introduced. These were extremely fast and had good overall performance. They failed to gain acceptance, however, because to ensure good performance they required considerable basic maintenance and so they were not cost-effective. Maintenance time can also reduce the time that an instrument is in effective use.

An economic advantage of automation, which falls outside the scope of Eqs. 1.1 and 1.2, lies in the potential silent-hours usage of a proven automatic method. Thus, a further two- to threefold increase in throughput for the analysis in question is possible and this is particularly appealing to a laboratory wanting to increase its analytical capacity without increasing its staff numbers. However, if silent-hours working is contemplated, several factors must be carefully evaluated. Safety in operation becomes paramount and the effects of power failure, blockage of reagent lines, and mechanical failures must be assessed and steps taken to ensure that no hazard will result if they occur. Designers of automatic analysers recognize the importance of continuous and unattended operation, and fail-safe circuitry is incorporated whenever possible. Nevertheless, it is not advisable to introduce silent-hour usage until the operator is fully conversant with the effects of mal-operation during the normal working day. As well as guarding against possible laboratory accidents arising from unattended operation of an automatic analyser, it is important to ensure that instrument failure does not result in irrevocable loss of sample results. It is essential that major faults result in automatic shutdown of the instrument. Ruggedness of the chemistry of the method is also important. Fluctuations in line voltage and ambient temperature can occur overnight, and if the method is sensitive to such effects, then incorrect results can occur even when the instrument is working satisfactorily. This knowledge is held by analytical chemists and serves to emphasize one aspect of their changed role when manual methods are superseded by automatic ones.

Silent-hours operation, which is commonly termed 'hands-off' analysis, requires the automatic analysis to operate to a set protocol. For a fully automatic instrument to run in this manner, it will require a feedback system comparing the results with check calibration standards. A calibration graph can be constructed from the analytical data, and the precision of this graph is easily evaluated. As the analyses proceed, the system can be monitored by reference to the check calibration standards. Should the performance remain within specification, the analyses can safely go on. The automatic instrument can then operate within the set protocols throughout the silent hours, taking full account of any variations in the instrument and its operating parameters.

For an automatic method to be preferable purely on economic grounds, its cost must be less than the manual cost by at least an amount equal to the cost of the automatic equipment amortized over a period of three to five years. For many of the more expensive instruments, particularly those in the clinical market, leasing agreements are common; and in these cases the annual cost must be less for the automated regime. However, this simple algebraic treatment is very approximate and takes no account of the differences in reagent costs, power requirements, and supervisory cost between the two methods.

In automatic continuous-flow analysers, reagent costs may well be significant. However, in many situations, suitable modification of the chemistry or improvements in design, such as transfer from Technicon AAI to AAII methodology, can produce savings in reagent costs. Porter and Sawyer [4] modified the chemistry involved in the analysis of total sugar in beverages by introducing a redox electrode system. This has two advantages over the original colorimetric system: it reduces reagent consumption and it reduces analysis time.

The economic treatment discussed so far is limited to analytical laboratories where samples are received from an outside source; it will not apply to laboratories attached to processing plants performing quality-control analyses. The cost of the automatic equipment, in these cases, will be small in relation to the plant cost, and it will be the improved precision of analysis and speed of response that will have the greatest economic significance. Automatic analysers in production lines are ideal for quality control, and there is ample scope for additional automation. However, this is an area where the arbitrary boundaries that have been introduced into analytical chemistry have hindered progress. Chemists are insular in their approach and would greatly benefit if they broadened their horizons and attempted to transfer technology from related areas to solve problems in their own fields.

Extreme care must be taken in estimating the cost of changing to automatic analysis. The manner in which the role and attitude of the analytical chemist must be revised have already been touched upon. There is a corresponding impact on the laboratory management and organization. To become fully conversant with an automatic analyser requires an understanding of the principles underlying the analyser design: these are likely to include electronics, mechanics and data processing. While a relatively unskilled operator can use the analyser when it is working satisfactorily, it is usually necessary to have more highly qualified staff available to detect, diagnose, and rectify faults.

The introduction of automation on an appreciable scale in a laboratory which has hitherto depended on manual analysis is almost certain to call for an adjustment of staff expertise, with an increased emphasis on non-chemical support disciplines. Staffing for the routine operation of the instruments may well be reduced to provide some cost advantage, but if the analytical laboratory is part of a multidisciplinary establishment, the ancillary needs can be met, at least partly, by internal staffing rearrangement. However, the effect on a small laboratory, staffed mainly by chemists, could be considerable, and this needs to be considered and planned for in advance of installation.

1.2 EXAMPLES OF SUCCESSFUL AUTOMATION

Reasons for automating include the reduction of staffing in the laboratory; the value of the end result; and the speed of response required.

In planning the introduction of automation into an analytical laboratory, it is important not only to consider all stages of the analysis, but also the wider context within which the laboratory serves the organization of which it is a part. Examples of laboratories that have engaged in effective planning are the Laboratory of the Government Chemist, London; the Nutrients Composition Laboratory at USDA, Beltsvillle, Maryland, USA; and Shell Development Company, Seal Hollow, Houston, Texas, USA.

1.2.1 Laboratory of the Government Chemist (LGC)

The LGC (see also Chapter 3) is unique in the UK and carries out analytical work for a number of government departments. Broadly speaking, the LGC has four major functions: (a) to carry out routine analysis for other government departments, (b) to advise those departments on chemical matters, (c) to carry out statutory obligations such as alcohol analysis in beers, wines and spirits and (d) to carry out research into analytical chemistry to support these other functions. These functions are directed in part by UK laws, but increasingly by directives from the European Commission in Brussels. Some of the tasks of the LGC are outlined in Table 1.1.

Table 1.1   Main reasons for the analytical tasks performed by the LGC

(a)		Tariff classification
(b)		Revenue collection or protection
(c)		Environmental control
(d)		Drug abuse
(e)		Surveys for public information
(f)		Statutory obligations

A typical requirement might be the analysis of representative samples of cigarettes to illustrate to the public the relative risks of the individual brands according to their dry tar content.

Because action is taken on the basis of these measurements, the data have to be precise and fully validated. Often, many samples are often involved, and, in addition, the continual pressure on staff numbers has justified a considerable investment in the introduction of automation and in defining a sensible strategy for training and management of automation skills within the laboratory.

In addition, the wide disparity of the sample matrices and the measurements required has meant that the LGC has produced a number of systems to automate a range of analytical problems. The value of automation to the LGC was investigated in the 1970s by a multi-disciplinary team. This took the form initially of adapting existing commercially available instrumentation to the various tasks in hand. As progress was made and systems introduced, a cost-benefit evaluation was carried out on all projects. When the savings were related to tangible benefits, such as better quality of results, attempts were made to convert them into cash terms.

Table 1.2 gives some of the reasons for the LGC setting up its automation team. The primary motivation was economic. LGC was often subject to constraints on staffing in parallel with large increases in analytical commitments. The introduction of cost-effective analyses, using mechanical or automatic instruments, reduces staff involvement and allows well-qualified people to be released for the development of new analytical requirements. The analysis of beer samples by multi-channel continuous flow analyser [5,6,7] and the introduction of a mechanical solvent extraction and identification system to analyse and measure levels of quinizarin in gas oil, both for duty purposes, were prime examples [8]. Both systems involved commercially available components and/or instruments integrated with modules designed and built in-house.

Initially, four people at the LGC were devoted to automation: three chemists and a physicist who had a background in electronics. As well as being trained in analytical chemistry, the physicist's skills in electronics were extended by a day-release course. Once major instruments were installed and in routine use, the resulting staff savings were then made available to the automation team to increase its effectiveness.

Table 1.2   Reasons for setting up an automation team

Economic
- Speed of analysis (surveys)
- Release staff for other duties
- Consistency of results

Requirements are complex
- Solutions not obvious
- Not met by instrument companies, for example GC

Specification
- Design considerations
- Involvement of chemist

Who operates?

Maintenance

Continued evolution of automated system

As the influence of the group expanded, further skills were added to it and at its height it numbered 26 people. The structure of the group has been described previously [9]. In simple terms it comprised three groups with responsibilities in electronics, computing and the analytical aspects of automation. It functioned by close co-operation between these groups the members of which had a good respect for each others' skills and a clear understanding of the customer laboratory requirements.

Speed and improved quality are also important. In the analysis of quinizarin in gas oil, transfer to the automatic regime immediately improved performance. Manual solvent extraction is a very boring task. An analyst, in an attempt to relieve the monotony, will set up a series of extractions in parallel. However, the sodium salt of the quinizarin is an unstable complex. Inevitably, variable times are taken for the solvent extraction, which leads to variable product development and imprecise results. Automation accurately sets the time for the extraction, removes this area of variability and provides consistent and reliable results.

Another fundamental reason for setting up a skilled team to review automation requirements is the complex nature of the analytical problems encountered. Often a more rudimentary analytical procedure precisely controlled by an automatic machine will be preferable to automating procedures like filtration and solvent extraction, which can be quite difficult. In this sense, the use of ion exchange columns for concentrating lead in drinking water, described by Bysouth *et al.* [10-12], is a more readily automated process than complex solvent extraction procedures. The latter may appear to be easy to automate, but when real samples are encountered, differentiating the organic/aqueous interface can be very difficult and impossible to control reliably. Many automatic instruments based on ion-exchange columns are now becoming commercially available. Few solvent extraction systems remain, even though some were sold in the late 1970s.

Where the problems of automation are more specifically related to the role of the laboratory, it is impossible for commercial companies to be able to offer cost-effective solutions on a one-off basis. When the LGC's Automation Group was at its height in the 1970s, there was little incentive for a company to develop equipment especially for the LGC. To overcome this, the Automation Group had to fill the gap, and the skills of mechanical engineering, electronics and software were important. The author's background experience at the Philips Research Group was important at this stage in terms of accurately specifying what could be done by hardware or software, or explaining how chemical procedures could be modified to suit the requirement.

Defining the specification of the analytical requirement and its solution are difficult and the analytical chemist's experience is vital. Only he or she can accurately predict the ruggedness or vagueness of analytical procedures. Customers know how they would like the automatic instrument to operate and will have a good understanding of the chemistry involved. How a specification should be drawn up is explained in Chapter 2, but unless it is carried out properly the equipment will be over- or under-specified; or if the problem is not correctly addressed, the automatic equipment will fall into disuse because it fails to achieve the overall objective.

When an automatic instrument is specified, it is important to know who will be operating the equipment. A good operator will run the equipment reliably and will give good results on a routine basis. However, if the operator is bored and does not keep all the equipment housekeeping in order, results will become unreliable. Most operators can cope with small problems, but if a complex problem occurs then an automation group skilled in all the various facets of automation may better be able to provide a prompt solution, corrective action, and get the instrument online again. Downtime can be extremely expensive and a properly maintained system will operate more reliably. Such a feature is in-built when an automation group has designed the system in-house.

Table 1.3   Aim of Automation Division

Good automation
        - at appropriate level of sophistication

Total systems approach
        - from sample preparation to reporting
        - under analyst control

Engineered version simple to operate

The final advantage of an in-house team is the ability to modify equipment to meet the changing needs of the organization. Experience gained from one automation project can be transferred to new areas. Analysing samples of beer for alcohol content at the LGC resulted in a fundamental change in the design of the flash distillation system [13]. When the development of a vapour generator system for the determination of selenium in water samples was required, this presented problems of design. This technique was originally proposed by Goulden and Brooksbank [14] and involved continuous flow digestion procedures coupled to vapour generation techniques. These procedures were very similar to the gas/liquid separator used for beer analysis. This saved a considerable time in the system development.

The aims of the automation group at LGC were very clear and are shown in Table 1.3. Simplicity was considered to be the best approach, with the minimum number of processes being involved. A more complex approach has many more chances of failure. The total systems approach is defined in Chapter 3. Essentially, it sets out to cover all aspects of the analytical process as defined in Fig. 1.2. It provides for the quality checks at operator, supervisor and managerial levels, and reliable results transferred in a readily digestible format. The Tar and Nicotine Survey described by Stockwell and Copeland [15] is a good example of the approach. The total process, from the statistical sampling pattern through to quality-controlled data, leads in its final format to results tabulated for public information.

Equipment held together by string and sealing-wax may have an appeal for the eccentric inventor but it has no place in a routine laboratory. A well-engineered instrument which is simple to operate, maintain and use will be of considerable advantage. It is also vital to have a fully documented and easy-to-read manual for the user and for service. Reliance on the designer's memory to fix a software or hardware problem some years after the equipment was installed must not be an option, even when in-house design teams are at hand. Commitment from the team to the project and from the user to the automation team are vital ingredients making any automation project successful.

1.2.2   Nutrients Composition Laboratory, USDA

Experience gained by a review in 1978 of the United States Department of Agriculture (USDA), Nutrients Composition Laboratory, in Beltsville, Maryland, USA, shows a somewhat different perspective, although many of the reasons are of equal value. Essentially, the Laboratory had to analyse a wide range of foods and to provide data on the nutrient content of these foods for public information. Like the LGC, it was involved in the design of appropriate sampling programmes. In order to meet these various demands, the automation team at USDA was set up and this, in parallel with a group headed by Professor Ruzicka and co-workers, led to the development of the techniques which became known as flow-injection analysis [16,17].

At USDA, a major requirement related to the production of large quantities of data and therefore rapid procedures needed to be designed. These had to be flexible and able to be redefined for new and different tasks. Resources were relatively limited and the technology already in use in the laboratory, i.e. HPLC equipment and auto-analyser modules, were used in the emerging continuous non-segmented flow analyses. A series of simple, but effective, detection systems to monitor those first reactions were built in consultation with small and adaptable local instrument companies. The enthusiasm of the group, and the need to solve real analytical problems, led to a series of patentable inventions on which several instrumentation companies have been founded. The group quickly and correctly noted that analytical methods need not reach a steady state, as in the continuous-flow regime described by Skeggs [18].

The reasons outlined for resorting to the use of automation were in many senses similar to LGC's motivations. However, the top priority was the ability of automation to provide greater precision of analysis. Cost was also important, since funds in any government agency are always under pressure. While USDA had a range of staff, it did not have the ability to recruit large numbers of additional people to carry out surveys. Automatic systems involving dangerous chemicals can, if properly designed, be carried out without risk and the safety aspect was also high on the list of priorities. A major incentive was simply the fact that, in the 1970s, automatic analysis was a new idea, one which could guarantee enthusiasm from the group employed at USDA. Their enthusiasm bore fruit in respect of the novel developments of flow-injection analysers.

USDA required that its automatic systems provided improved accuracy and precision. The type of work directed that a very high sample throughput was available and that there was a need for a flexible approach to the measurements. Since the staff already had a good background and experience of high pressure liquid chromatography applications and the use of Technicon AutoAnalyzers, it was not surprising that the flow-injection system designed by this group used HPLC pumps and Technicon samplers. The system was made in a very simple fashion which was easy to assemble and repair by the staff in-house. Whereas many groups working in automation were willing to take away the chart recorder trace and rely on computer data processing of results, USDA continued to use analogue output. It provided a trace of all of the data and in the event of any problems with samples, clearly indicated what went wrong, i.e. it indicated any disaster. Replacing this simple output of software would have been expensive and not as flexible as the trained analyst reviewing the shape and form of the peaks produced. Not being content with developing a new approach to automation in the laboratory, the USDA also invested considerable effort in defining the theory of the technique so it fully understood both the theory and practice of the methodology developed [19].

1.2.3 Shell Development Company

The Shell Development Company in Seal Hollow, Houston, Texas, USA, has been investigating robotic systems as part of its continuing programme of automation (Taylor et al. [20]). Shell is a good example of an industry committed to automation.

Figure 1.2 shows the trends in analytical science which have encouraged Shell and others to invest in this area. Analytical chemistry is being revolutionized as a result of the introduction of computers; in addition, the industry is facing increasing competition and new demands of environmental legislation.

Shell has found many cost-effective solutions using robots. The future requirements and trends for their systems are as follows (these, of course, apply to many organizations):

1. At present, most of Shell's robots are self-correcting in that they can be programmed to respond to particular kinds of error. Newer robots will be more involved in making on-line decisions, and could ultimately evolve into expert systems.
2. It would be a natural development to integrate robots with the logistics of the sample request system adopted in Shell and to record their output automatically in central data archives.
3. Walk-up capability could be provided for customers by robots which could read procedures and sample designations from bar codes and perform and report the corresponding analyses.
4. The use of robotic facilities could improve reproducibility between different laboratories and tighter controls could be developed. So information from robots will probably be increasingly used in regulatory standardized tests, such as those required to meet environmental regulations.

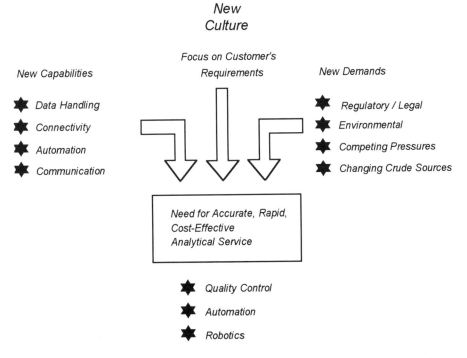

Fig. 1.2 Trends in analytical science which affect the needs for laboratory automation. *Reproduced with the permission of the copyright holder, Zymark Corporation.*

Finally, as robot users develop more familiarity with the capabilities of systems, manufacturers are conceiving and requesting more and more complex applications. Vendors may prefer to select less complex, high-volume applications which may have greater application volume potential. Between these two aspirations, there may develop a presently unfulfilled need for some third party to bridge the gap between what the users would like to have and what the vendors would like to provide.

1.3 EFFECTS OF AUTOMATIC ANALYSIS

The early years in the development of automated analytical techniques were largely devoted to solving the scientific and technological problems of designing and constructing equipment capable of performing analytical procedures in a precise and repeatable manner, with sufficient reliability to allow it to fulfil its task with minimal supervision. During this phase, little thought was directed to the wider problems which confront a laboratory which changes from a regime based on manual analysis to one dependent on automation. However, many laboratories now have several years of experience with automated systems, and it is possible to address the broader issues. Brief consideration is given here to the problems which the introduction of automated analysis pose for the analytical chemist and for the laboratory management.

1.4 THE ANALYTICAL CHEMIST

In recent years a large number of experienced analytical chemists have witnessed the introduction of automatic techniques into their laboratories; indeed, they may have helped to initiate the changeover in the interest of efficiency. As the dependence on manual methods of analysis is reduced, analysts have become aware that the demand for manipulative skills and experimental judgement is being eroded. At first sight, the role appears to have been degraded in that the accumulated skill and expertise are replaced by a requirement merely to operate an instrument, which is largely incapable of responsive judgement. Equally, they feel less able to defend the results produced by the automatic instrument with the same personal conviction that they can apply to their manual efforts. Closer examination reveals that the function of analysts has not been downgraded; in a laboratory committed to automatic analysis they hold a central and indispensable role. However, the revised role requires adaptation and some changes in attitude, notably in taking a much broader view of analysis than hitherto. Analytical chemists retain ultimate responsibility for the status and quality of the results produced by the machine, and it is imperative that, with the full co-operation of the manufacturer or in-house designer, they acquire an understanding of the principles and operation of the machine. They will come to appreciate that, if the automatic instrument is working to specification in all respects, then the limiting factor in the quality of the results is usually the reliability of the chemistry of the method being used. Thus analysts have not conceded responsibility for the methods used; rather, they can exploit their analytical knowledge to modify the chemistry to make it more compatible with the instrument facility and more resilient to minor variations in sample composition, e.g. pH in order to offset, in part, the lack of responsiveness of the treatment to such changes.

The analyst has gained another responsibility, that of specifying the requirements for, and performance of, any new automatic equipment to be purchased or constructed, with particular reference to stages or parameters where close control is critical. This is, in general, a new and unfamiliar task for analytical chemists, and they can only discharge it effectively with the active collaboration of colleagues experienced in other relevant scientific disciplines. The alternative, which can be fraught with pitfalls, is to accept manufacturers' literature at face value. The importance of devising a proper specification for the analytical requirement is further stressed in Chapter 2.

In summary, the introduction of automation into an analytical laboratory leads the analytical chemist, unwillingly in some instances, away from experimental individualism towards being part of a team in which the origin of the samples and the requirements of their originator assume an importance equal to the analysis itself.

1.5 LABORATORY MANAGEMENT

Laboratory managers must be conscious of the changed role of the analyst as a result of automation and endeavour to provide the facilities and motivation to ensure that analysts adapt productively to their new responsibilities. The management team should encourage a rapport with instrument designers and manufacturers and ensure that advantage is taken of appropriate educational and training facilities.

Taylor *et al.* [20] have paid considerable attention to the changing role of management, particularly with regard to the introduction of robotic systems. Management has the responsibility to define the best areas for automation in terms of cost-effectiveness and probability of success. Without adequate resources, a project will fail. Laboratory managers must act as the interface between senior management and analysts to ensure that the appropriate incentives, resources and education are made available.

Failures need to be clearly identified so that the lessons learnt can be widely communicated. The recognition of tasks well conceived and resolved is of importance, and management must be seen to give credit where it is due and should take the responsibility themselves for any failures. Ensuring that the group is exposed to new technologies and to other groups working in the area, so that good lines of communications are set up, can also be a priority.

REFERENCES

[1] International Federation of Clinical Chemists (IFCC), Centre du Medicament, Universite de Nancy 1, Nancy, France.
[2] Stockwell, P.B. and Godden R.G., *Journal of Analytical Atomic Spectrometry*, 1989, 4, 301.
[3] *Yorkshire Water Methods of Analysis Handbook*, 1988.
[4] Porter, D.G. and Sawyer, R., *Analyst*, 1972, 97, 569.
[5] Sawyer, R. and Dixon, E.J., *Analyst*, 1968, 93, 669.
[6] Sawyer, R. and Dixon, E.J., *Analyst*, 1968, 93, 680.
[7] Sawyer, R., Dixon, E.J., Lidzey, R.G. and Stockwell, P.B., *Analyst*, 1970, 95, 957.
[8] Tucker, K.B.E., Sawyer, R. and Stockwell, P.B., *Analyst*, 1970, 95, 730.
[9] Stockwell, P.B., *Journal of Automatic Chemistry*, 1979, 1, 216.
[10] Bysouth, S.R., Tyson, J.F. and Stockwell, P.B., *Analyst*, 1990, 115, 571.
[11] Bysouth, S.R., Tyson, J.F. and Stockwell, P.B., *Journal of Automatic Chemistry*, 1989, 11, 36.
[12] Bysouth, S.R., Tyson, J.F. and Stockwell, P.B., *Analytica Chimica Acta*, 1988, 214, 318.

References

[13] Lidzey R.G., Sawyer, R. and Stockwell, P.B., *Laboratory Practice*, 1971, 20, 213.
[14] Goulden, P.D. and Brooksbank P., *Analytical Chemistry*, 1974, 46, 1431.
[15] Stockwell, P.B. and Copeland, G.K.E., in *Topics in Automatic Chemical Analysis*, Edited by Stockwell, P.B. and Foreman, J.K., Ellis Horwood, 1979.
[16] Stewart, K.L., Beecher, G.R. and Howe, P.E., *Analytical Biochemistry*, 1976, 79, 162.
[17] Ruzicka, J. and Hansen, E.H., *Analytica Chimica Acta*, 1975, 78, 31.
[18] Skeggs, L.T., *American Journal of Clinical Pathology*, 1957, 28, 311.
[19] Vanderslice, J.T. and Beecher, G.R., *Talanta*, 1985, 32, 334.
[20] Taylor, G.L., Smith, T.R. and Kamla, G.J., *Journal of Automatic Chemistry*, 1991, 13, 3.

2

Principles of automatic analysis

2.1 INTRODUCTION

Chapter 1 discussed the reasons for automation and attempted to set out the needs of analysis. This second chapter defines the objectives of automatic analysis and describes how, in principle, these can be achieved.

Analytical procedure is a systems problem and the sampling, pretreatment, measurement, data collection and reduction, and final reporting all have to be considered in a fully automatic approach. Computerization is often considered to be synonymous with automation but, although microprocessor technology is certainly changing the face of automatic instrumentation and influences both the control aspects and the data reduction, computerization is only a part of automation. Computers should simply be considered as 'tools of the trade' within the area of automation.

Automation has been applied for a number of years in process control instrumentation, but the major impetus to introduce automatic devices into laboratories stems from three sources: (1) the introduction of the continuous-flow principles as outlined by Skeggs [1]; (2) the general demand for clinical chemical measurements, which represents a ready and sizeable market for instrument companies, and, more importantly, (3) the ability to handle large volumes of data and package them in a form suitable for presentation to analysts and customers, through the use of mini- and micro-computer systems linked to a control computer.

The availability, at a reasonable price, of computer power, along with the various associated peripherals that are imperative for a viable computer system, is seen as being the most important influence on future growth in this area. Recent years have seen the decline of the mainframe due to the increased power and availability of the personal computer. Today's Pentium-based IBM-compatible PCs with 100 MHz internal clocks and 500 megabyte hard disks are a far cry from those that were popular with chemists a decade ago.

This needs to be balanced against the ever-increasing cost of reliable and effective software that is also required to maximize the usefulness of computer hardware. Recent years have seen the increased popularity of the Microsoft Windows operating system. MS Windows is an attempt to give PCs the window management environment of professional workstations. It provides an easy-to-use graphical interface, and shelves are groaning under the weight of applications for analytical chemistry. A major problem is that the jargon surrounding computing presents a formidable barrier for inexperienced users. MS Windows alleviates part of the problem by providing a consistent package-to-package user interface. However, the complexity of popular PC platforms and the programming languages themselves only serve to obscure issues that were once transparent to the inexperienced programmer.

Two options are available:

1. The tailoring of an off-the-shelf data acquisition package to meet your needs. Example: National Instrument LabVIEW for Windows. LabVIEW is a graphical programming system for data acquisition and control, data analysis and data presentation. The developer is able to build 'virtual instruments' to acquire data from plug-in boards presenting results through graphical user interfaces. It provides an innovative approach and an alternative to text based programming.

2. The development of dedicated systems using a low-, medium- or high-level programming language, the most common being PASCAL, FORTRAN, BASIC, C, and Fourth Generation languages. One of the more popular approaches at the moment is that of object-orientated programming. This is a technique which allows the programmer to 'picture' concepts as a variety of objects, each with their own attributes and functionality.

Rapid progress towards more effective automation will require a concerted effort in developing the necessary software. It is necessary to select the appropriate tool in order to exploit the power of existing hardware.

The availability of computing power has had a significant effect on analytical instrumentation generally. In the early days of automation, the most significant impact was achieved by applying first mini-computers and subsequently microcomputers to instrumentation. For many of the major analytical instruments, nuclear magnetic resonance spectrometers, mass spectrometers and inductively coupled plasma spectrometers, the addition of the latest computer technology provides inherent flexibility and ease-of-use for the analyst. The benefits of this computer power will be outlined here by reference to the Finnegan Mat Quadrupole Spectrometers. However, transferring these advantages to other techniques is equally valid, but it is not thought advantageous to repeat the descriptions.

The Finnegan single stage quadrupole powers a very flexible system for several analytical problems. The user has a choice of electron ionization positive chemical ionization or negative chemical ionization. The Quadrupole can be linked to gas or liquid chromatography and several direct insertion probes. The data features powerful, multi-layered software using high performance workstations, that make data analysis and instrument control easy.

Key features of the software include:

1. A graphical user interface with a pop-up Windows environment.
2. Menu and mouse driven functions for data analysis.
3. Extensive on-line help with context-specific instructions.
4. A key word search feature which assists with unfamiliar terms.

Using a high-performance, multi-tasking data system, the user can call up and work in several windows on-screen simultaneously. For example, the user can set parameters for the next sample run, view a chromatogram or spectrum and perform background acquisition of data at the same time.

Both new and experienced users can take advantage of software to lead them through instrument set-up and data acquisition. Automated quantitation routines are available to calculate compound amounts in samples. These routines offer complete flexibility to generate custom report formats. By integrating retention time information and internal or external calibration curves can be calculated. Quantitation data can be exported to popular spreadsheet programs.

An Instrument Control Language (ICL^{TM}) gives both power and flexibility in operating the system. With this operators have complete control to create or customize instrument procedures, and design experiments to exactly meet laboratory needs. The instrument can optimize analysis by making real-time decisions during data acquisitions.

For example, based on incoming data the analyst can:

1. Change probe temperature, or other acquisition parameters.
2. Choose unlimited selected ion monitoring (SIM) segments.
3. Switch among many experiments such as source fragmentation in API, positive and negative ion analysis, and all MS/MS options.

In addition, optical applications software enables the user to capture screens, label figures and import text and graphics into reports, documents or other formats.

Mass spectrometers provide control over both GC and LC autosamplers for complete system automation. For maximum throughput and productivity, procedures can be set in advance and downloaded to the GC or LC at a later stage - an ideal feature for laboratories that require 24-hour operation.

Independent control of all analytical parameters means that non-repetitive analyses can be automated. The computer software can also lead a new operator through an experiment on the instrument.

If a laboratory needs to share information across a base of users, a standard Ethernet^{TM} network is provided. This allows file sharing between systems. PCs, Macintosh computers and other workstations can be used as extra terminals. Networking allows the user to:

1. Submit data to other departments.
2. Share data with anyone on the network.
3. Share hardware resources such as printers and modems.

Most importantly, the networking capabilities can direct data and information immediately to meet the need for timely analytical results.

Four serial (RS232) ports are provided for flexible communications to other instruments, i.e. autosamplers. In addition, five analogue inputs, two analogue outputs and 3 TTL input/outputs are provided to ensure complete flexibility. These allow auxiliary instruments under direct control and the ability to process data generated by these in real time. These features extend the range of facilities on these expensive but worthwhile analytical techniques. There is no doubt that future developments in computing will continue to have a radical impact on these instrumental systems.

Because of the importance of the advances made in the field of clinical chemistry, instrument companies have tended to concentrate on this market. There is, however, another, and even more exciting, market in industrial process control. In this area there are real problems of sample handling and sample matrix effects. In contrast, the clinical market is fortunate in that its problems relate mainly to blood and urine, which are readily overcome compared with the problems experienced in the industrial area. This is not to say that blood and serum do not present significant matrix effects, but the number of customers with the same and/or similar problems is very large. In industrial chemistry the matrix problems vary within the industry and within the customer base. A significant problem in clinical chemistry, however, which has repercussions on the other market sectors, is that it takes a long time for an instrument to be designed, developed, evaluated and accepted into use. The time scale for this is of the order of 10 years. This puts considerable constraints on instrument development and only the most successful instrument companies can afford the financial investment necessary. Indeed, many companies have attempted and failed to break into this market, often investing very large sums without producing a successful instrument.

There are signs that companies are becoming increasingly aware of the industrial market and some attempts have been made to develop a systematic approach to this problem. Whereas in clinical chemistry the matrix is usually blood or urine, in the industrial area there are many varied matrices. The volume of sales for any matrix is often insufficient to justify the development investment required. An alternative philosophy is needed to meet the requirements economically. The Mettler range of automatic instruments provides one example of a systematic approach to automate a range of analysers. More recently the Zymark Corporation (Zymark Center, Hopkinton, Massachusetts, USA), in the introduction of its Benchmate products, has defined procedures which can be tailored to individual laboratory needs by using essentially similar modules. These modules are co-ordinated with a simplified robotic arm. Several tailor-made systems have been developed which have a wide appeal and are easily configurable to particular needs.

Automatic chemistry draws on a whole range of disciplines. Therefore the student of automation, or the systems designer, must be willing to use technology developed for achievements in one sector and to apply it to another area, which is often only loosely related to the first. There are therefore problems of education, communication and specification to consider. The area upon which automatic chemistry impinges is vast and it is therefore impossible to provide comprehensive coverage in this Chapter. An attempt will be made to highlight some of the more interesting and more recent avenues of development.

2.2   EDUCATION, SPECIFICATION AND COMMUNICATION

These three aspects present many problems in automatic analysis. They are considered separately here, but, of course, they are highly interactive.

2.2.1   Education

For the laboratory with analytical needs which can be met by purchasing a commercial automatic analyser and using it strictly in accordance with the maker's instructions, methods and servicing, the training needs of the operators can usually be met through a course provided by the manufacturer. Most major manufacturers offer training at their own premises on a regular or as-required basis. But for those laboratories with a large and varied workload, commercial automatic analysers are likely to be viewed as facilities to be modified to meet new needs. Frequently, the chemical methods needed will be developed by the laboratory staff, and, in these circumstances, it is essential that several of the staff are familiar with component design and performance, sample-transport mechanisms, data processing, and control technology. For these people the education and training requirements are more demanding. At present, these cannot be met in a wholly satisfactory manner. The underlying training requirement is to create an understanding not only of the relevant scientific principles, but also of the philosophies of automation, which are evolving rapidly as the range of applicability of automatic analysis continues to grow.

To educational establishments, the teaching of automatic analysis poses three problems. It is an interdisciplinary subject involving engineering, electronics, and computing in addition to chemistry. Equipment for demonstration purposes is expensive and at the present time, very few teaching staff have more than a rudimentary experience of the subject. Nevertheless, colleges are generally aware of the problems and have made important, although limited, attempts to overcome them. For example Betteridge *et al.* [2] have described a microprocessor-controlled automatic titrator which is of value as a teaching aid, and the relatively inexpensive, and simple flow-injection technique seems ideal for the same purpose. In the USA, Malmstadt *et al.* [3] have studied the teaching aspects for some years and have developed a modular experimental approach. Teaching manuals and equipment for this are available for purchase.

In the UK, automatic analysis is taught in a number of postgraduate M.Sc. and diploma courses devoted to instrumental analysis. With increasing experience and availability of equipment, such courses should become more effective in presenting the principles of the subject. However, the important philosophical aspects would be better presented by experienced workers in the field who can pass on not only the practice but the thinking that generates progress in automatic analysis. There is ample scope for specially designed short courses, with the use of equipment loaned by manufacturers to illustrate development. It is a little sad, therefore, to see that as more of the instrumentation being used in degree courses becomes automated in various degrees, the academic staff appear to be relying on this as being adequate training for the subject. Clearly, since the new generation is not being exposed to the failures in automation, but only the limited successes, there is an even greater need to provide the necessary background and education.

Some limited progress is being made, such as the discussion and workshop sessions on automatic analysis now featured at the Pittsburgh Conference on Analytical Chemistry and Applied Spectroscopy [4], and at least one region of the American Chemical Society has sponsored training schools in the subject.

In the late 1970s, Betteridge, Porter and Stockwell [5] organized, under the auspices of the Royal Society of Chemistry, a series of summer schools on automatic analysis at University College, Swansea, Wales, UK, which drew together world experts to present the foundation lectures and programmes. This series of lectures covered the various facets of automation, including management and economics, in addition to developments in philosophy and techniques. Techniques such as thin-film chemistry, the use of microcomputer-controlled systems and flow-injection analysis were discussed in some depth. Table 2.1 illustrates the scope of the lecture programme. In addition, a range of tutorial sessions catering for the participants' own particular interests was arranged. Tutors comprised the lecturing staff and local experts. Twenty practical experiments, made available by a number of instrument companies and other institutions, enabled the course members to gain first-hand experience of a wide range of instruments. One novel feature of the course was a tutorial session in which the course members had to develop a strategy for finding an automated solution to an important analytical problem. This problem related to the analysis of cigarette smoke to determine its tar, nicotine and carbon monoxide levels. Small groups considered the problem first and then combined into three large groups, each of which prepared a joint presentation and discussed the development strategy. At the end of the course each of these three groups made a formal presentation. The actual approach taken to solve the problem was then presented, and this is described in Chapter 3.

Table 2.1   Organization of summer schools (Betteridge *et al.* [5])

Form of teaching	Time allocated	The following areas of expertise were involved
Lectures	16 x 45 min = 12 hr	Instructors: academic, industrial, governmental and a degree of clinical experience.
Tutorials	(a) Course problem 3 x 45 min (b) Specialist 4 x 45 min Total 5¼ hr	Real world automation problems discussed along with criteria for solution.
Practicals	(a) Fixed exercises 4 x 1½ hr (b) Free choice 2 x 1 hr + Wed. afternoon	Instrumentation companies allowed hands-on experience with full range of equipment from single modules to complete systems.
Residential Course	Open all hours	Relaxed atmosphere and direct access to a range of worldwide experts at minimal costs.

The various approaches were then discussed and contrasted by the lecturers and course members. This problem and the surrounding discussion served to illustrate the various philosophies and principles outlined by the course staff throughout the course. The response to this course was particularly enthusiastic, and whilst there are undoubtedly improvements and modifications that could usefully be made in any future venture, the success of the course proved that the basic formula was correct, and, above all, that this type of course is needed in the UK and, no doubt, elsewhere. Attendance, however, became limited and the course has been stopped. Despite the levels of robotics, automation and computerization being introduced into laboratories no other course has been set up which adequately provides the necessary background to the subject to allow a new generation of systems designers to become adequately trained. Generally speaking, most training is carried out in-house by users or directly by the instrument companies. In addition, Zymark Corporation has followed the lead of Technicon Instruments and has organized specific user meetings. At these annual meetings, users can present the latest developments that they have made and mix on an academic and social level with other users. The instrument company also can present its latest developments as well as listening to its customer base to identify possible new products.

2.2.2   Specification

To derive the full benefit from automatic equipment it is essential that proper consideration is given, at the outset, to specifying the analytical requirement in sufficient detail to ensure that the characteristics of the equipment installed match the analytical needs as closely as possible. This is a major new role for the analytical chemist and one for which laboratory management, manufacturers, and educational institutions must develop, on a continuing basis, principles for guidance. Specification is of fundamental importance where automatic analysers are to be designed and constructed rather than purchased from commercial suppliers. Over-design is expensive and time-consuming; under-design can lead to the inability of the product to meet the full analytical requirement, a frustration which often cannot easily be eliminated by later modifications. Specification includes the chemical procedure to be used. Direct conversion of a manual method for automatic operation may not always prove satisfactory, especially if the manual method includes steps, such as precipitation, which do not conveniently lend themselves to automation. In such circumstances it is often preferable to modify the method so that the full economic and scientific benefits of automation can be incorporated. Considerable advantages can be gained if such a study is undertaken, even if the result is that new manual procedures are introduced rather than automatic ones. A worrying trend in clinical chemistry is that the necessity for automated methods far exceeds the requirement for specific or precise analyses. Several instrumental methods have been introduced which are difficult to calibrate, especially in multi-analyte analyses.

If proper attention is given to calibration and standardization, a well-designed, constructed and maintained automatic analyser will operate reproducibly over long periods in the hands of a trained operator.

As the dependence on manual methods of analysis is reduced, the analyst becomes conscious that the demand for his manipulative skill and experimental judgements is being eroded. The changing role of the analyst was discussed in detail in Chapter 1.

The analyst gains additional responsibilities for specifying the requirements for, and performance of, any new automatic equipment to be purchased or constructed, with particular reference to stages or parameters where close control is critical.

This is, in general, a new and unfamiliar task for analytical chemists, and they can only discharge it effectively with the active collaboration of colleagues experienced in other relevant scientific disciplines. The alternative, which can be fraught with pitfalls, is to accept manufacturers' literature at face value. The importance of devising a proper specification for the analytical requirement cannot be over-stressed. It is essential, however, to see clearly the implications of automation in its widest sense and to consider the analytical calculating and reporting procedures from the outset. The analytical requirements must be clearly defined and may not be the same as those of the corresponding manual procedure.

The introduction of automation into an analytical laboratory leads the analytical chemist away from experimental individualism towards a managerial team-consciousness in which the origin of the samples and the requirements of their originator explicitly assume an importance equal to the analysis itself.

Figure 2.1 attempts to illustrate the various facets of designing and developing an automatic analytical system. From this it can be seen that the overall solution is most likely to be a compromise between chemistry and computer hardware and software. It can be seen that for any particular problem all the necessary technology could be purchased directly from a commercial company. Equally, all the development could be done 'in house', although the most realistic solution is likely to be a hybrid of commercial modules and software and home-made devices, with some of the software modified to suit the particular needs of the laboratory. The role of the laboratory within the organization of which it is a part has an overriding influence on the specification. Whilst the specification and production of a device is a multidisciplinary task, analysts have a vital role and they alone can properly define the constraints, be they chemical, statutory or legal. In clinical laboratories the overriding requirement is the speed of analysis, whereas the precise result may be of secondary importance. In legislative requirements, it may be most important to have consistent results with those obtained by statutory methods. In-house construction should not be undertaken lightly and Porter and Stockwell [6] and Carlyle [7] have discussed these problems in some depth. Providing an acurate asessment of costs in design and development is important, since if these are excessive then the benefits of automation will be lost.

Analysis of analytical requirement in such detail may at first sight seem daunting, but the investment in time will be of great value. Embarking on such an exercise should not necessarily end in an automated instrument - it is perfectly possible that the end result is that the analyses are discontinued owing to limited value in the results. However, in the main, a compromise but easy-to-use automatic solution will result which will be used reliably on a routine basis. One of the major questions that needs to be resolved is how in general terms the analyses should be carried out. Some of the available strategies are outlined later in this chapter.

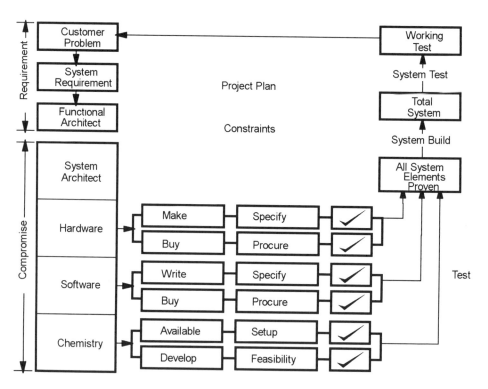

Fig. 2.1   A systematic approach to solving problems of laboratory automation.

2.2.3 Communications

From the above it is obvious that a systems designer for laboratory automation must be conversant with and able to utilize a whole range of disciplines, from chemistry to electronics, and from computer software to statistics and instrument design. The users of the automation, their managers and the customer for the analytical results also have a contribution to make in deciding the specification.

Very often the user with little or no experience of automation defines his requirements in terms that he can understand: this is often not the best approach. The technical literature tends to present the view that existing instrumentation can and does solve any problem the user might have. At the specification stage it is necessary to establish the needs for the instrumentation and the solution to the problem will then be a compromise between the economic and scientific requirements meeting the needs of the end-user. The solution will also attempt to make use of the technology and resources available to the organization. Unfortunately, there is a problem of communication, because every discipline, despite the apparent belief that only computing suffers from this problem, has its own jargon behind which the specialists hide. This represents a barrier to the chemist. However, given a willingness to overcome this problem, it is quite possible to make an assessment of the abilities of each technique.

Few books attempt to cover the area of automatic chemistry. Technical papers which relate to automation have most often been presented in analytical journals dealing with the basic subject area, and the details of the automation are less well documented than the chemistry. Aspects of management, education and economics are given scant treatment despite their importance. The *Journal of Automatic Chemistry* [8] was launched to fill this void in the literature and experience has shown that there is a wealth of technical experience awaiting publication. In addition, it is vital for the suppliers of automatic instrumentation and computer systems to set up adequate lines of communications with their customers. User groups are a valuable asset so that experiences, good and bad, are shared and any problems resolved. A user who becomes frustrated for lack of support or information will become a source of bad reference. A satisfied user, on the other hand, will provide valuable input which may benefit future users and assist the company in providing instruments that the user needs.

2.3 DEVELOPMENTS IN THE PHILOSOPHY OF AUTOMATION

The overriding benefits of automation are economic in nature. The rapid introduction of automation into clinical laboratories and the large market open to potential manufacturers have prompted many developments in instrumentation. Far less consideration has been given to the philosophical approaches to the subject. Until recently these could be considered in three categories: (1) the approach in which methodologies are adapted to the techniques available such as in continuous-flow analysis; (2) a total systems approach, such as that developed at the Laboratory of the Government Chemist, which attempts to define the total analytical requirement, including transmission of the result to the customer and which uses available techniques to solve the problem in the most effective manner; and (3) the approach described by Arndt and Werder [9] in which a systems study is made of the available manual procedures, which are then broken down into a number of steps, a complete rationalization of which then generates a flexible range of equipment designs to meet the varied needs of an analytical laboratory. While the first approach typifies that followed by the majority of the instrument companies, the second is obviously the province of the systems designer working in a multidisciplinary laboratory. The third approach offers a fresh and encouraging input from one of the world's major instrument companies embarking on an automation programme in concert with an industrial chemical company already committed to the principles and benefits of automating their work pattern.

Many of the developments in automatic chemistry, including the introduction of continuous-flow analysis and radioimmunoassay, have evolved from the clinical environment. One exciting development, which offers a fourth approach to automation, is that of thin-layer, or solid-phase, chemistry. The chemistry is carried out on dry-to-touch, separate but interacting layers on a supporting membrane, and has been described by Curie *et al.* [10] and Spayd *et al.* [11]. The technology, developed for photographic processes, has been perfected to the point where up to 16 different layers can be produced with a high degree of precision. Although photographic principles are not directly involved, the technology has been transferred to the analysis of blood and serum. The details of the technique are discussed in later sections of this chapter, but, in principle, a small section of film, termed a chip, is used to transport the sample and to perform the essential chemistry through a series of layers designed for the appropriate analysis. Photographic film of high quality has been in production for a number years and the precision attained in the thin-layer approach to clinical applications can be attributed to the experience of first black-and-white and then colour photography.

2.4 PRINCIPLES OF AUTOMATIC INSTRUMENTS

Besides understanding the philosophy and concepts of automatic instrumentation, it is necessary to make a clear economic assessment of the advantages to be gained from the introduction of automation. There can be no substitute for practical experience in solving such problems. The field is wide, so a complete review is not practical. Some developments are described here to stimulate the reader into deeper research. It is important to evaluate the various developments that have become available through both large and small instrumentation companies.

In basic terms, all automated analyses of samples in the liquid state are performed by one of two methods, discrete or continuous, and occasionally by a combination of the two. There are several basic subdivisions in each group and some examples will be described here and throughout the other chapters of the book to provide a starting platform for the systems designer.

In the discrete method each sample is maintained as a separate entity, it is placed in a separate receptacle, and the analytical stages of dilution, reagent addition and mixing are performed separately by mechanically transporting the sample to dispensing units where controlled additions are made to each sample individually. Likewise each treated sample is presented in turn to the measuring unit (colorimeter, electrodes). In continuous analysis the sample is converted into a flowing stream by a pumping system and the necessary reagent additions are made by continuous pumping and merging of the sample and reagent streams. Finally the treated sample is pumped to a flow-through measuring unit and thence to waste.

In the design of an automated analytical instrument, or selection of the most appropriate commercially available type, the choice between the discrete and continuous approach is fundamental. Both types are considered in detail under the various technique headings, but the broad relative merits of the two approaches are set out here for general reference.

The discrete method has the advantage that samples can be processed at a high rate. For example, commercial colorimetric analysers are capable of yielding between 100 and 300 measurements per hour, whereas for continuous analysers a processing rate of 20-80 samples per hour is normal. However, the high-throughput discrete analysers are appreciably more expensive than the continuous analysers.

Because the discrete approach retains the sample as an entity, cross-contamination between samples is almost entirely eliminated. Since the sample retains its identity throughout the analysis, its fate at any time is known and there is little chance of confusing one sample with another at the measurement and recording stage. In continuous analysis the identity of the sample is lost once the processing is under way and in a normal continuous analyser several samples are being processed at any one time between the sampling and measurement stages. Because of the regular timing intervals between stages, as controlled by the pumping rate and length of delay lines, there is usually no difficulty in associating each recorder response with a particular sample, and, indeed, it is becoming increasingly common to include a sample identification system on the recorder trace. Nevertheless, in simpler instruments devoid of this facility, problems can arise if a succession of samples have zero responses. Where this situation is suspected the insertion of frequent standards and quality control materials in the sample sequence affords regular datum points. More significantly, unless precautions are taken, interaction between successive samples can occur in a continuous system, thereby causing overlap and loss of discrimination at the recording stage. In general, interaction can be minimized by optimizing the design of the timing sequence between samples and by reducing the processing rate and, uneconomically, by inserting water between each sample. The higher the sample throughput, the greater is the interaction between samples and this accounts for the upper limit on sample processing rate in continuous analysers.

Continuous analysers have the merit of being mechanically simpler than the discrete ones. Sample transport and reagent addition require only a suitable pump. Peristaltic pumps are almost always used in commercial continuous analysers and the ready availability of multichannel peristaltic pumps enables a single motor to control the entire sequence of events. Discrete analysers require a number of moving parts for sample transport, and valves or automatic syringes for reagent addition. In general, therefore, discrete analysers are more demanding on maintenance than the continuous type and regular attention to the moving parts is essential. Clearly this requirement can by minimized by good design and optimum selection of materials of construction.

In the discrete method each analytical operation requires a separate group of moving parts, and, in general, discrete analysers are best suited to heavy loads of chemically straightforward analyses. For this reason their major field of application has so far been in hospitals for clinical analysis of blood and urine, using a multi-chemical system capable of several simultaneous simple analyses where the stages involved are dilution and addition of one or two reagents followed by measurements. Discrete solvent extraction units have been designed and proved, but await commercial exploitation. Continuous analysers are favoured where chemical pretreatment stages are necessary. The design problems involved in incorporating pretreatment stages in a flowing system are less formidable than with discrete operation and many satisfactory techniques are now available. Consequently, continuous analysis has some advantages where intermediate separation stages are required between sampling and final measurement.

Continuous analysis requires flexible tubes which are not attacked by the materials under examination, and this places certain limitations on the scope of the method. Certain reactive and corrosive materials cannot be satisfactorily pumped, although advances have been made in the development of inert plastics and other synthetic materials. Displacement pumping with the aid of a liquid compatible with sample and reagents provides an alternative, though generally inconvenient, approach. No such limitations arise in discrete analysers because there is no restriction on the choice of materials for sample and reagent containment.

The comments above regarding the relative merits and limitations of the discrete and continuous methods are intended as broad generalizations only, although they will frequently be of value in deciding the approach to a new problem. The two approaches are not mutually exclusive. For example, a number of commercially available automatic amino-acid analysers utilize continuous ion-exchange separation followed by discrete analysis of individual fractions of column eluate. As emphasized earlier, it is the chemistry of the method which determines the instrumental approach; only in relatively few cases, of which clinical analysis is the principal example, has the choice of instrumental approach been optimized. In other instances the chemist is likely to find that, even though many papers have been written on automatic analysis, the subject is still in its infancy. This is one of the main reasons why a multi-disciplinary approach to automation commends itself as being the most profitable means of answering the first and most vital question: how should a new problem in automatic analysis be approached?

2.4.1 Discrete systems

Some examples of discrete analysers are outlined here. When these were first introduced, materials technology was not able to cope with the constraints put on them by the instruments designed. However, progress on materials has been significant over the last two decades, making this approach to automation more viable. However, many lessons can be learnt by the various automation concepts introduced. The major problem with these designs is aptly illustrated by reference to the Vickers Multichannel 300. This had become so complicated that it has been discontinued; it was also designed to be so fast that a major bottleneck occurred in simply loading the samples. A further drawback was the degree of maintenance that had to be carried out even on a routine basis.

Vickers Ltd designed and developed the Multichannel 300 analyser primarily to meet the need for detailed analysis of blood samples. It is a multi-channel instrument, the standard model being provided with 6, 8 or 12 individual channels, though alternative configurations are possible because once a sample is received each channel functions as a separate unit. The instrument was conceived to discharge high workloads such as would be met in large hospitals or in institutions concerned with health-screening of the general population. In theory, there are few ultimate limitations to the number of channels which could be incorporated. Apart from considerations of space, complexity and cost, the more general limitations are likely to be the number of subsamples which can be taken from the initial sample and the number of methods which have been fully proven for automatic operation. With the exception of a central sample-distribution line, there are no common facilities in the chemical stages of the analyser, each channel having its own individual reagent-addition and photometric-measurement units. Since each channel is capable of processing 300 samples per hour, a 12-channel instrument is theoretically able to yield 3600 results in every hour. In an analyser of this type, the reliability and reproducibility of individual components is of the utmost importance in order to minimize potentially high maintenance costs. Nevertheless, the approach is advantageous in that a breakdown of one channel is confined to that channel and the results from others are unaffected.

Before individual analyses are commenced, the blood samples, contained in labelled closed rectangular vials, are centrifuged and then loaded in magazines. Unequivocal coding of samples is of paramount importance in the processing of the number of samples of which the Multichannel 300 is capable; a 12-digit system is used, six for the patient's name and six for departmental identification. The sample-loading mechanism holds several magazines. The vials are delivered from the magazine to the next stage by a system of pawls which opens a gate and allow the vial, held against the gate by a constant-tension spring, to be pushed out. As soon as a magazine is emptied the next one is automatically indexed into position. An optical sensing technique rejects vials which are underfilled or which contain haemolysed or lipaemic samples. Rejected samples are discharged to separate vial holders. The rejection system can be manually over-ridden, so that a sample that has been designated unsatisfactory can be used, if necessary. Samples accepted by the checking procedure move to the primary sampling and dilution stage. The sampling-probe unit transfers a volume of sample, together with diluent, into an open-topped vessel. The vertical positioning of the sampling head is controlled by a cam; when it is lowered the probe breaks the vial seal and draws in sample. The sampling head is in two parts in the form of a rotary valve. The lower part, which remains stationary, is connected to the supply of diluent. The upper part carries two capillary probes and rotates in half-turns. Thus one probe dips into a vial to withdraw sample while the second is delivering sample, followed by diluent, taken from the previous vial. The holders for diluted sample are contained in a distribution assembly which serves the individual analysis channels. When the diluted sample reaches an analysis channel, termed a reaction console, a secondary sampler and diluter removes an aliquot into a cavity in a thermostated reaction rotor. All further treatment is performed in the cavity. Each rotor contains 60 such cavities, or 120 if blank facilities are required. The inner surface of the cavities is non-wetting, and chemically inert. The time-sequencing is such that a complete rotation of the rotor takes 12 minutes. During rotation each cavity passes reagent dispensers situated round the periphery; the geometric positioning of the dispensers depends on the analysis being performed and the incubation times required. At the final sample position is situated a suction probe which transfers the sample into the colorimeter cuvette (or to the atomizer of a flame photometer). The cavity is subsequently washed free from any adhering residue and dried in readiness for the next sample.

The colorimeter provides double-beam operation so that simultaneous blank corrections can be carried out. This eliminates errors due to variations in the light-source emission and sample matrix effects. The double beam is obtained from a single lamp by using a beam-splitting prism. Results are presented on a teletype, together with a punched paper tape which can be used for further data processing. Results are presented accompanied by the vial code which is obtained by moving the sample vial to a reader which detects the digits impressed on the label.

A far simpler example of the discrete or batch principle is the Anton Paar (Kärntner Strasse 322, A 8054 Graz, Austria) VAO digestion device. It consists of a rail transport system which lifts racks of samples from a loading area through five zones of heating or digestion and then to a waiting station for unloading. A schematic diagram of the system is shown in Fig. 2.2. The maximum temperature in the digestion zones can be set anywhere between $0°$ to $400°C$. The actual temperature in each of the five heating zones can be adjusted by settings which determine how deeply the vessels are inserted into each of the heating zones. The digestion period can be set anywhere between 10 minutes and five hours. Therefore, with a total digestion period of 30 minutes through five heating zones, a rack of 40 samples would be completed every 6 minutes, yielding a throughput of 40 samples per hour. The simple mechanical device offers reliable operation and conformity to a range of protocols, for example Environmental Protection Agency, USA, and Contract Laboratory Procedures. All samples are treated in an exactly similar manner with little operator attention, with unskilled operators. Additionally, a hydrogen peroxide attachment can be added to dispense automatically dispense the addition of this reagent to further extend the range of applications of the technique.

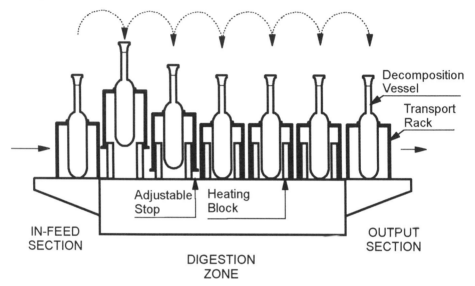

Fig. 2.2   The Anton Paar VAO digestion device.

The VAO can be programmed for a specific decomposition by adjusting the following parameters:

Selection of digestion vessels: the size of the digestion vessels depends on the sample volume. Standard vessels are 25, 40, 50, 100 and 150 ml glass. For extreme trace analysis, quartz and Teflon vessels are available. Vessels can also be supplied with dust covers and reflux extensions.

Selection of digestion time: By means of the time switch, at the left front corner, the time interval per digestion zone can be set at from 2 to 60 minutes. Multiplying this by five gives the total digestion time for transporting the vessels through the VAO.

Selection of temperature profile: The maximum temperature required for a digestion is selected with the thermostat to the right of the timer switch. However, for many digestions it is necessary first to heat the sample slowly and sometimes even to continue heating throughout the cycle. For this reason, each heating block has a set of pegs which can be fixed at eight different heights, designated from bottom to top with the numbers 0 to 7. The trays of samples sit on these pegs and thus permit the sample vessels to be more or less immersed into a particular heating zone.

For a given application, the peg positions may be as coded 7/6/4/0/0 meaning that in the first zone, the pegs are located at the top in position seven for least heating; in the second zone in position six for slightly increased heating; at the third zone in position four, and at the fourth and fifth zones the vessels would be immersed completely, without any pegs, for maximum heating.

Therefore, a programme for a particular digestion might be indicated as:

1. Digestion vessels: 40 ml glass, long neck.
2. Advance interval: 7 minutes.
3. Maximum temperature: 270°C.
4. Temperature profile: 7/6/4/0/0.

With this programme, the vessels reach a maximum temperature of 270°C at zone four. The total digestion time would be 35 minutes. Applications include measurement of total recoverable metals in waste water and the measurement of trace metals in air filters.

2.4.2 Recent discrete analysers

Whilst the reliability of performance of the earlier models was often in doubt the same cannot be said for those instruments available in the 1990s. Of these, the Olympus AU5000 Series is a good example; it integrates high processing capacity with good reliability and is highly economical. Reagent volumes are minimized and only the most reliable materials are used for the system components, lowering the overall operating costs.

The precise configuration can be chosen according to the laboratory workload. Table 2.2 clearly illustrates the capacities of the various models. These instruments will support a wide range of analytical methods including end-point assays at one or two wavelengths and with one or two steps as well as rate analyses.

The instrumental range is designed primarily for the clinical laboratories but 10 different wavelengths between 340 nm and 800 nm can be selected to match the test requirements. Sixteen photometric sensors are available. Serum quality data, including lipernia, icterus and remolysis, can be obtained in addition to the serum blank. All of this information can be obtained without sacrificing the number of channels available for chemical analysis.

Table 2.2 Capacities of various models in the AU5000 Series

Model	No. of test items	Samples/ hour	Test results/ hour
AU5211	16	165	2,640
AU5221	16	330	5,280
AU5223	32	165	5,280
AU5231	24	330	7,920

The basic configuration of the AU5000 Series is shown in Fig. 2.3.

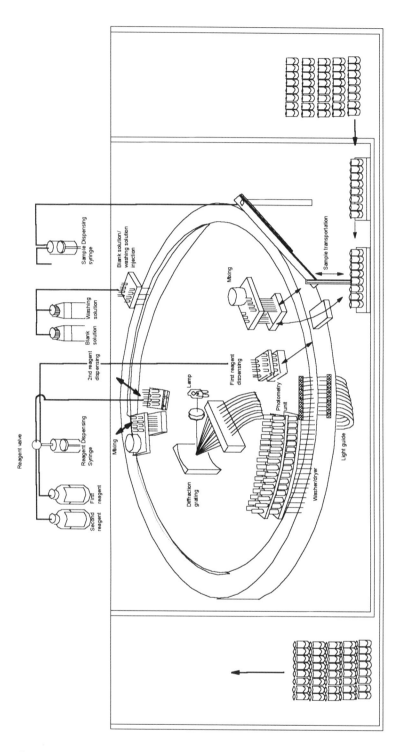

Fig 2.3   The basic configuration of the Olympus AU5000 Series discrete analyser.

Samples are held in racks on the right-hand side of the instrument. Colour coding is provided for automatic rack identification - white for routine samples, red for emergency (stat) samples, blue for reagent absorbance measurements, yellow for calibrations, green for quality control samples and orange for repeat samples. This speeds the identification of priority emergency samples. Bar coding of these racks eliminates identification errors, allowing racks to be placed in any order. Samples for repeat analyses can be added indiscriminately. When the bar code option is active, quality control samples can be analysed along with routine samples in the same rack. As a result, quality control becomes more consistent without sacrificing processing capacity.

The sample probe's vertical stroke can be as long as 100 mm and this, with the in-built liquid level sensor, allows a wide range of primary tubes and centrifuge tubes to be held on the trays. Sample racks are automatically transported in the front section of the analyser, which in turn simplifies maintenance should the nozzle become blocked or if mechanical problems arise. The AU5000 Series provides precise sample dispensing capability. Each of the two sample probes aspirates a volume sufficient for a maximum of four chemistries per sample and dispenses it into eight cuvettes on the twin reactor lines. The probes are constructed from water-resistant plastic which minimizes any dilution from wash solution or contamination. The built-in level sensor limits the immersion of the probe in the sample. The inside and outside of the probe as well as the level sensor are rinsed after each use to further prevent contamination.

Reagents are held in a refrigerated compartment which accommodates a wide range of containers with up to a 2 litre capacity. Liquid level sensors are provided to indicate reagent shortage. With the AU5000 Series, a single syringe is used to dispense the reagent. Ceramic valves are used to prevent measurement errors and to improve durability. The reagent line is automatically fed with wash solution to bleach the various dispensing lines. This reduces maintenance and provides consistent results. The reagent line configuration is minimized such that the reagent volume usage is approximately half that of other systems.

The AU5000 Series uses a dry bath system for cuvette incubation. Keeping the cuvettes out of direct contact with the liquid makes the washing of cuvette exteriors unnecessary. After the analysis is completed, the AU5000 Series injects washing solution into all of the cuvettes automatically, to cleanse the cuvette interior of impurities. During analysis, cuvettes are always kept clean with a thorough washing system, assuring highly reliable data generation. Twin reaction lines provde a compact system configuration for the system. The pre-spectrophotometry system uses a diffraction grating to divide the light from a single lamp into 10 wavelengths which are then measured photometrically through optical fibres. This eliminates strong irradiation falling on the sample reagent mixture, so that bilrubin and other basic substances are protected from deterioration. A single lamp is used for photometry of both the inside and outside reaction lines. This means that no difference is observed in light intensity. This is shown in Fig. 2.4.

The extensive data check functions provided on the AU5000 together with individual channel quality control checks ensure that the data consistency from this instrument range is extremely high. If the measured data falls outside the reagent optical density (O.D.), normal, repeat run, panic value, dynamic or other specified value ranges, abnormal data will be marked with warning flags. When the analytical procedure is completed, reaction curves for up to 3200 tests are displayed on the CRT for data confirmation by the analyst. Screen display of the latest QC sample and reagent blank data provide quick and clear information and if they should fall outside the pre-determined range, an alarm is activated and warning flags appear on the screen. The meticulous quality control for each channel eliminates data inconsistency and simplifies data management.

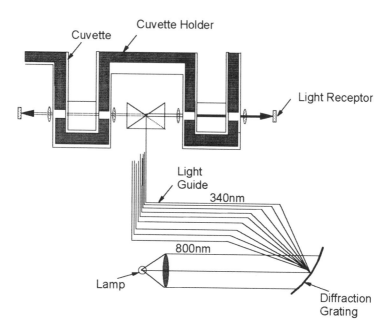

Fig. 2.4   Lamp configuration for AU5000 Series.

2.4.3   GeMSAEC and DACOS parallel processing

The automatic analysers described so far in this chapter process a batch of samples sequentially. The GeMSAEC system (General Medical Sciences - Atomic Energy Commission) developed at Oak Ridge National Laboratory, Tennessee, USA is designed for very fast rates of processing of clinical samples and treats a batch of samples in parallel. It utilizes centrifugal force to mix solutions and transfer them to cuvette rotors which intercept the light-beam of a photometer as they rotate [12].

Figure 2.5 shows the basic principle of GeMSAEC. The sample and reagent holders are fabricated from Teflon, as is the cuvette rotor in which the photometric measurement is made. The design of the indentations for holding sample and reagents is such that no mixing occurs when the apparatus is stationary but on spinning the rotor at 500-1500 rpm complete mixing occurs within 10-15 s, and the mixture is transferred to the cuvette. The equipment is designed to handle small volumes of sample and reagents, usually of 0.2-0.3 ml.

In early studies a 15-cuvette rotor was used, but, subsequently a 42-cuvette rotor has been designed [13]. At the high speeds of rotation each sample is photometrically measured over a very short period of time, about 0.05 s. Therefore if one cuvette contains a reagent blank solution a simultaneous double-beam analysis of 41 samples is obtained. Allowing for complete mixing of sample and reagents to occur, the analytical results on a sample batch can be obtained some 30 s after the start of the experiment. Drainage and wash-out of the rotor take a further 90 s. The rotor unit can then be removed and replaced by another one already loaded with sample and reagents.

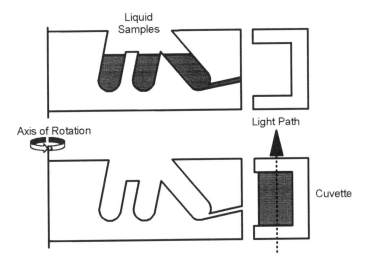

Fig. 2.5   The GeMSAEC automatic colorimeter. *Reproduced with permission of Anderson [12].*

The GeMSAEC analyser can be operated either with premeasured volumes of sample and reagent, or with unmeasured but sufficient volumes. In the first case, standard automatic micropipettes can be used to load the rotors. In the second case the equipment must be modified to enable a fixed final volume to be measured and transferred to the cuvette - one way of achieving this is shown in Fig. 2.6. When the rotor is spun, the solution is forced into a transfer tube and any excess of liquid is drained away; the measuring tube is then mechanically turned through 180° and the measured volume delivered to the cuvette. Alternatively, a series of siphons can be used to measure and transfer solutions as shown in Fig. 2.7.

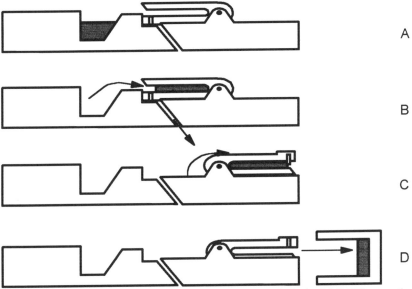

Fig. 2.6   Centrifugal fixed volume sampling and transfer in the GeMSAEC automatic colorimeter. *Reproduced with permission of Anderson [12].*

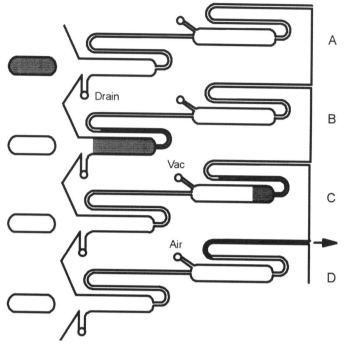

Fig. 2.7 The GeMSAEC automatic colorimeter: siphon method of fixed volume sampling and transfer. *Reproduced with permission of Anderson [12].*

Since all the cuvettes intercept the light-beam over a period of about 0.05 s, the results can conveniently be displayed and recorded to provide a permanent record. For the workloads for which GeMSAEC is designed, computer processing of results is clearly justified and interfacing of the system with a computer has been described [14]. The ability of GeMSAEC to produce readings within seconds of the initiation of a reaction makes it ideally suited for the study of kinetic systems. With the aid of suitable software [15], plots of extent of reaction as a function of time can be generated and the approach has been successfully applied to enzyme kinetic studies [16].

The major disadvantage of the GeMSAEC is that although several analyses are run in parallel, the system is discontinuous and the rotor has to be stopped for reloading; it is also difficult to carry out more than one type of assay in a rotor at any one time. This problem was overcome in the DACOS approach (Discrete Analyser with Continuous Optical Scanning) described by Snook *et al.* [17]. In this approach, reaction tubes are situated at the periphery of the rotor, which turns in discrete steps, a few degrees at a time, to enable tubes to be loaded with sample and subsequently washed when measurements are complete.

Instead of the tubes rotating rapidly through the light-beam, the beam itself rotates on the same axis as the rotor. The signal pattern from this type of analysis is therefore identical to that from the conventional centrifugal analysis approach. Long reaction times can be accommodated either by slowing the motion of the rotor, or by suppressing the washing, sample, re-transfer and reagent addition phases for one or more cycles; the scanning speed can also be varied within wide limits.

Figure 2.8 shows the general arrangement of the instrument with 100 optical cuvettes radially dispersed around the vertical axis of the reaction rotor. The optical components of the dual-channel monochromator are rigidly mounted within a machined aluminium housing (which is not shown in the diagram), and the housing itself is solidly fixed to the top of the rotor.

The precision optical shaft encoder accurately relates the signals from the photomultiplier with the appropriate cuvettes. The shaft encoder tube, which contains the reflecting prism, is driven by a small synchronous motor and associated gear train, which provides the rotation of the scanning beam. The photomultiplier and the collecting lenses of the detector optics are also mounted rigidly on the rotor, so that the relative positions of all the optical components, except for the rotating reflecting prisms, are rigidly fixed with respect to each other.

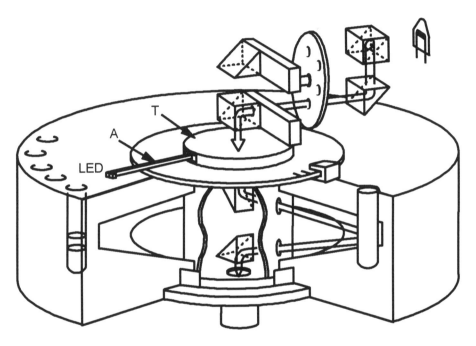

Fig. 2.8   Schematic diagram of discrete analyser with continuous optical scanning.

The cuvette-cleaning, sample-transfer and reagent-addition mechanism remains static relative to the mainframe of the equipment. The analytical rotor is moved through $3.6°$ with respect to the equipment mainframe every 24 s. Since the rotor carries within itself the continuously scanning beam of the absorptiometer, another monitoring device is required to define the scanning sector, and to relate the effect of the double movement of the beam to the mainframe. This is achieved by fitting a light-emitting diode (LED) to the end of the arm A, which is fixed to the rotating encoder tube T.

A light-shield (not shown), drilled with 100 holes on its circumference, covers this rotating miniature light-source, and the passage of the light behind the 100 holes is monitored by two photodetectors fixed to the equipment mainframe. The detectors are not adjustable and always define the maximum reading sector. This low-resolution encoder provides the interrupts necessary to define the START and END of the reading sector. The computer software routines recognize these interrupts and relates the data to the appropriate cuvettes, thus avoiding any mismatching of the data recorded.

The flexibility required to handle procedures where a lag phase is needed, or where reagents must be added, is incorporated into the software routines. In this manner, any data points which do not meet the specified criteria for the analysis are ignored. Stepping of the reaction rotor to receive a new sample is achieved by a simple rim drive arrangement which is activated when the scanning beam leaves the reading sector on the twelfth scan. Because of the large yield of analytical data it is necessary to integrate a computer into the overall design, chiefly for data handling but also in part to function as a process controller. Although Snook *et al.* [17] used a Texas 960A computer, microcomputer systems would also be suitable.

The classification of kinetic methods proposed by Pardue [18] is adopted in the software philosophy. The defined objective of measurement in the system is to obtain the best regression fit to a minimum of 10 data points, taken over either a fixed time (i.e. the maximum time for slow reactions) or variable time (for reactions complete in less than 34 min, which is the maximum practical observation time). In an analytical system generating information at the rate of 50 datum points per second, with reactions being monitored for up to 2040 s, effective data-reduction is of prime importance. To reduce this large quantity of analytical data to more manageable proportions, an algorithm was devised to optimize the time-base of the measurements for each individual specimen.

The calculation routines for the regression slope and its standard error are time-shared with the data-acquisition, before the cuvette is cleaned ready to receive a new sample at the solution-transfer position. When the scanning beam leaves the reading sector other computer-controlled tasks begin. The printing of results is also time-shared with the mathematics packages. The system is very flexible and overcomes many of the objections to other conventional centrifugal analysers; it can monitor reactions which take place over times ranging from 18 s to 45 min. The principles embodied in the system described by Snook *et al.* [17] have been incorporated into a new instrument marketed by Coulter Scientific Inc., Florida, USA [15].

2.4.4 Integrated robotic centrifugal analyser

The Cobas Fara II represents a logical extension of the centrifugal analyser by integrating a robotic sampling pretreatment system into the design. It is one of the most qualified systems for routine and special chemistry and offers multi-assay automation and flexibility. Six measurement modes are available; absorbance, fluorescence polarization, fluorescence intensity, nephelometry, turbidimetry and ion selective electrodes. These modes of operation enable the automation of almost any special chemistry routine measurements, including substrates, enzymes, therapeutic drug monitoring, toxicology, lipids, specific proteins, immunoassays, coagulation and electrolytes.

Figure 2.9 shows the instrumental design of the Cobas Fara II. The versatile rack system is specifically designed to provide fast, simple reagent handling. The reagent rack allows testing with up to 75 on-board reagents using up to 11 reagents per single test. The rack can incorporate up to 160 samples, controls and/or standards. Binary coding on these racks is used to provide identification and to help minimize errors. The reagent racks can accommodate a wide variety of sample vials and reagent containers.

Two programmable, microprocessor-controlled, pipetting arms work independently of each other, enabling automated handling of samples and reagents. Programmable pipetting minimizes set-up time and ensures precise, on-line sample and reagent control. This reduces the possibility of errors caused by the reagent preparation in an off-line mode.

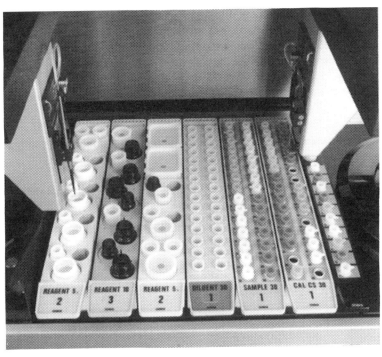

Fig. 2.9   The Cobas Fara II robotic centrifugal analyser. *Reproduced with permission of Roche Diagnostics.*

Sample and reagent volumes can be varied from 0 - 95 µl in 0.1 µl increments while larger volumes can be varied in 1 µl increments from 0 - 370 µl. These micro-volume capabilities are particularly important for paediatric, geriatric and animal testing. They offer considerable reagent cost savings.

The Cobas Fara II provides on-line sample pre-dilution or pre-treatment, enabling total automation of assays such as serum, proteins and urine chemistries. Pre-diluted samples can also be pre-incubated before transfer to the analyser. These on-line features minimize labour and errors while ensuring result integrity.

The system is continuously monitored to alert the operator to errors such as antigen excess, high background fluorescence readings, loss of sample/reagent integrity and deterioration of electrode slope range.

Fluid-level detectors which use capacitive measurement ensure sufficient reagent and sample volumes. These detectors are integrated into the reagent probe and sample tip and require no maintenance.

A comprehensive, on-board software package provides real-time, multi-tasking computing power. Real-time data retrieval allows easy access to calibration or quality control data and test results - including control and calibrator names and lot numbers, raw data and patient results - for storage, transmission, modification or deletion. The interactive keyboard and video monitor display are easy to operate and enable quick selection and viewing of data and parameters. The operator has complete control at all times and message prompting provides immediate direction. Routine and STAT worklists can be modified or deleted, even while the instrument is running. The 'Run Optimization' mode provides maximum throughput efficiency by combining the possibilities to run simultaneously different tests with the same reaction mode in one rotor and to set test priorities independently.

Special or routine chemistry runs can be interrupted at any time to perform rapid STATs with simultaneous multi-analyte analysis. The Cobas Fara II provides STAT test results in as little as 60 seconds with true 'Sample-Prioritized' capability. Routine testing resumes automatically from the point of STAT interruption.

Test results can be displayed and printed either by patient or test with cumulative storage. Patient data are automatically stored until specifically deleted. The 'Patient Identifier' function permits accurate sample labelling and identification.

Whilst specifically developed for the clinical market, the Cobas Fara II has many other areas of application, for example in environmental testing, food testing and animal toxicology.

Table 2.3 Outline of environmental test kits available on the Cobas Fara II instrument

Adaptations Available:

Analyte	Range	Analyte	Range
Alkalinity	(0 - 50 mg/l)	Magnesium	(0 - 30 mg/l)
	(0 - 250 mg/l)		
		Nitrate/	(0 - 2.0 mg/l)
Ammonia	(0 - 2.0 mg/l)	Nitrite	(0 - 10.0 mg/l)
	(0 - 20 mg/l)		
		Phosphate	(0 - 1.0 mg/l)
Bromide	(0 - 5.0 mg/l)	Silica	(0 - 25 mg/l)
		Sulphate	(0 - 50 mg/l)
Calcium	(0 - 25 mg/l)		(0 - 200 mg/l)
			(0 - 500 mg/l)
Chloride	(0 - 100 mg/l)		
	(0 - 500 mg/l)		
Iodide	(0 - 50 µg/l)		

Table 2.3 sets out the range of methods applicable to environmental testing. Tests on Environmental Protection Agency (EPA) reference materials showed that the results were statistically equivalent to the true values quoted. Other advantages claimed are that the number of samples that can be analysed in a given time period can be increased six-fold, developing and optimizing new procedures and improving existing ones is much easier, the volume of sample required is reduced by 90%, change-over time, from one chemical procedure to another is eliminated and no instrumental set-up equilibrium time is needed.

2.4.5 Automated single-test analysis system

The approach to laboratory automation described by Arndt and Werder [9] involved the close co-operation of a commercial instrument company with a laboratory attached to a large industrial chemical organization. The analytical workload in industrial wet-chemistry laboratories is characterized by a multitude of procedures and methods, as well as small numbers of samples in serial runs of analyses. This has been one of the reasons for the resistance in this field to automation. A new approach was needed and this has emerged as a modular subdivision of manual analytical procedures into basic operations. These are assigned execution parameters that determine the detailed operation. From this it is possible to derive a conceptual design of an automatic analysis system with which it is possible to run individual samples and small and large series of samples. Each sample is contained in its own cup, the automated units are self-cleaning, and existing analytical procedures may be used. A complete range of instruments based on these principles is available from Mettler AG, and instruments are easily tailored to suit an individual laboratory's requirements. Control is either by desk calculator, or for more sophisticated requirements, by a computer system controlling individual modules and co-ordinating overall control and reporting.

This approach represents a significant advance on the philosophy adopted by other instrument manufacturers. Once the utility of the approach has been proved in industrial laboratories, it will be interesting to see how far the philosophy will develop and solve problems directly, rather than simply mimicking manual procedures.

The instrumentation developed employs units for (1) automation of the basic operations; (2) sample transport; (3) central control; (4) the entry and weighing station; and (5) the output for results and logistics.

The heart of the system comprises the automatic units that carry out the basic operations. The individual steps of the operations are specified in a set of execution parameters, which are controlled and monitored by a microprocessor that also calculates the basic results. Apart from monitoring all operations for correct action, it is necessary to avoid the generation of incorrect analytical results and to signal the malfunctioning of any subsystem. The machine operation should allow the state of any procedure or module to be readily and easily visible.

The sample-transport mechanism is the physical link between the units for the basic operations and it moves the sample cups to the entry ports. The sample identification system ensures that samples are available to the appropriate unit at the right time. The mechanism functions like a railway system: it receives a command to move a cup containing a standard volume of sample from one place to another and then waits for the next instruction, which may require transport of the next sample cup or of the same sample to a different module.

Whereas the microprocessor controls an individual basic operation, the central computer, which has all the analytical procedures held in its memory, controls the particular analytical procedure required. At the appropriate time, the central computer transmits the relevant set of parameters to the corresponding units and provides the schedule for the sample-transport operation. All units are monitored to ensure proper functioning. If one of the units signals an error, a predetermined action, such as disposing of the sample, is taken. The basic results from the units are transferred to the central computer, the final results are calculated, and the report is passed to the output terminal. These results can also be transmitted to other data-processing equipment for administrative or management purposes. The central control is, therefore, the leading element in a hierarchy of processors. Figure 2.10 shows a schematic diagram of the system. The configuration of any particular system can be tailored to individual laboratory requirements. The entry and weighing station is the most important interface between man and machine. Brief comments arising from visual inspection of samples may be entered, and these will appear with the analytical results.

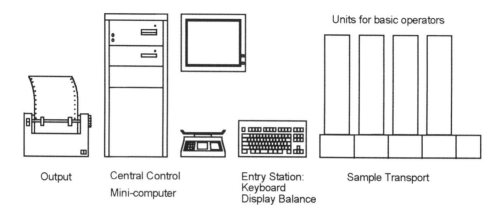

Fig. 2.10 Schematic representation of a system with several units for automation of basic operations, controlled by a mini-computer.

A sample of specified weight is normally required in the procedure. An interactive form of weighing is used, in which the display or printing unit of the entry station indicates whether or not the sample has been accepted. Before analysis it is necessary to specify the code number of the analytical method that is to be used, and to store this in the memory of the central control. To indicate where samples are located, it is necessary to identify them before weighing. Optical readers are therefore mounted on the sample-transport mechanism to register each sample. The sample is identified by a unique code placed on the outside of the sample cup.

After all of the primary input data have been entered by the operator, the sample is placed on the transport mechanism. Once the procedure is initiated, the remaining steps are carried out completely automatically.

The basic results from the individual units are processed and then combined to form the final result which is produced on the report printer. Results that deviate from an expected value by more than a preset tolerance may be marked or commented on. Additional information, such as sample identification and origin, is also made available. To ensure complete control by the analyst, the basic raw results may also be recorded in analogue form. Sample identification is provided so that the data can be re-analysed. Fully automatic systems require careful monitoring of the supply of reagents and the disposal of waste chemicals. To achieve this, fluid levels are monitored, and if they are low, an alarm signal is issued to the operator.

The whole system, with its internal and external interfaces, is designed so that it can be adapted to the needs of any particular analytical wet-chemistry laboratory, and so that responsibility for the analysis can be assigned to the operator. The system can be arranged in many ways and one of these, relating to the SR 10 system titrator, is shown in Fig. 2.11 (a) and (b). The samples enter the unit at the entry station and then the sample carrier moves the sample to the dispensing station where solvents or reagents are dispensed. The treated sample is then moved to one of four working stations, each of which can be fitted with a combination of a measuring and a reference electrode; it is also possible to attach a sensor for photometric titration. Up to six burette tips may be inserted at each working station.

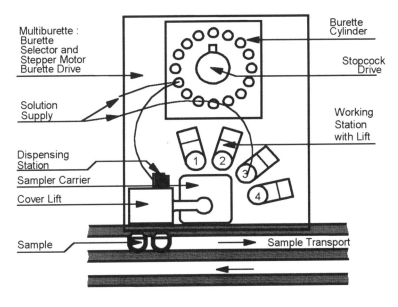

Fig. 2.11   (a) SR 10 System Titrator: arrangement of mechanical unit.

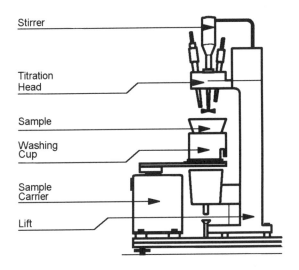

Fig. 2.11   (b) SR 10 System Titrator: working station with lift.

Before the sample is moved to the desired work station, the burette tips and the electrodes are rinsed, and the burette tips are primed to provide fresh solution. A washing cup that contains the conditioning solution for the electrodes while they are not in use is lowered and replaced by the sample cup.

The titration cycle starts with first a homogenizing period which allows dissolution of solid samples, flushing with an inert gas, or application of a chemical reaction. The sample may also be heated to a predetermined temperature. Next, a precise volume of reagent or reagents is dispensed if required. While the sample is being stirred a titration can be performed, either to a relative or an absolute end-point, or incrementally with or without equilibrium detection. Several titration modes are available, including potentiometric, amperometric, voltammetric and spectrophotometric.

The titration cycle, like most of the other functions, can be repeated at will. Back-titrations are therefore possible, as well as multiple titrations for multi-component analyses. At the end of the cycle, the sample is returned to the sample transport. All dispensing is from a multi-burette system with up to 20 dispensing assemblies, each with a total delivery volume of 10 or 20 ml.

The procedures are entered and stored in a conversational manner in a control computer. The software transmits the execution parameters to the titrator, accepts raw data, and calculates the final result. The analytical procedure to be used, with any auxiliary data, is first specified on the keyboard, then the sample is weighed (if necessary) and put on the transport; the rest of the operation is automatic. The four working stations permit configuration of the system for four different types of titration requiring different electrode combinations and reagents. The universal design permits performance of a range of functions which are specified by the user and depend on the hardware configuration (electrodes, reagents, amplifiers, options etc.) and the assay methods stored in the software. The accuracy of results, chemistry permitting, is better than 0.1%. Transfer of a manual process to the automatic regime takes between one and two hours. While this approach to laboratory automation is expensive, an organization with a large number of titration tasks can easily justify the investment: it can amortize the instrument over three or four years. The device operates at the combined speed of up to 10 skilled analysts and is probably more reliable.

2.4.6 Thin-film or solid-phase chemistry

Thin-film or solid-phase chemistry has been developed by Eastman Kodak. A brief explanation of the technique is given above. The salient principles of the technique can be seen by reference to the determination of urea in serum; Fig. 2.12 shows a schematic diagram of the 'chip' used and also illustrates the chemistry integrated into the multilayers. The spreading layer is an isotropically porous non-fibrous layer, 80% of which is void, and has a mean pore size of 1.5 μm. The spreading and metering action, which is aided by surfactants, compensates for any differences in sample size (10 μl applied) and serum viscosity. A constant volume per unit area is then naturally applied to subsequent layers of the chip. High molecular-weight materials, such as protein, are removed by this layer and consequently do not interfere with the subsequent analysis. Titanium oxide is incorporated in the spreading layer to improve its reflectivity. It also acts as a white background for reflectance measurements of the colour density produced in the reagent layer. The reactant layer contains the urease which catalyses the hydrolysis of urea in the sample to produce ammonia. Water from the added serum swells the gel, allowing the urease to diffuse into the layer. The layer is buffered to pH 7.8, which maintains the ammonia at a low level and consequently extends the range of the assay. A third layer consisting of cellulose acetate butyrate with selective permeability allows non-ionic materials such as ammonia and water to pass through to the indicator reagent, which in this example is N-propyl-4-(2,6-dinitro-4-chlorobenzyl) quinolinium ethanesulphonate. The free ammonia diffusing into this layer reacts with the indicator to form a dye which has a molar absorptivity of approximately 5000 $l.mole^{-1}.cm^{-1}$ and a broad absorption peak at 520 nm. The reflectance density is measured from the peak at 670 nm. The final layer is a clear polyester support upon which all the other layers are coated. It is transparent and allows measurement of the colour intensity of the compound formed in the indicator layer.

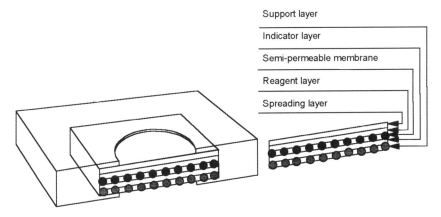

Fig. 2.12 Schematic diagram of the Eastman Kodak slide for solid-phase or thin-film chemistry, illustrating the chemistry integrated into the multilayers.

Performance trials and evaluation tests on the technique indicate that it is both reliable and accurate, and, in addition, that the specificity is sufficient to cope with most clinical requirements. An evaluation was made by Haeckel *et al.* [19]. If this approach is successful, the dispensers and tubes in laboratories will become redundant. It may well become possible for a clinical test to be undertaken close to the patient rather than in the laboratory. Whilst the techniques have as yet been used only for clinical analyses, there are many other potential applications, for example in the water industry. However, the very nature of the technique necessitates development by Eastman Kodak. Very few users will be able to influence the choice of analytical problems to be tackled by this unique approach.

2.5 AUTOMATIC CONTINUOUS ANALYSERS

Continuous automatic analysers (CAA) are characterized by the fact that the transport of samples and reagents along the system is effected by establishing a gas or liquid stream flowing through a manifold. The sample and reagent(s) are mixed in a number of ways, and a variety of intermediate operations including continuous separation units (dialysers, extractors etc.) can be involved in the operation of this type of analyser which also typically uses continuous detection in a flow-cell through which the stream carrying the reacting mixture is passed. The transient signals provided by the measurement system, whose time-dependence and shape are a function of the operational mode used, are analysed by a conventional recorder or a microprocessor. Some of the features of these signals (for example the peak height, width and area) are related to the analyte concentration.

These analysers resemble liquid and gas chromatographs, although their background is markedly different. They differ from discrete automatic analysers (DAA) in various respects, namely the fashion in which samples are transported and mixed with diluents and reagents, the manner in which carry-over between samples and reagents is avoided and the type of detection used.

The classification of automatic continuous methods is based on the way in which carry-over between samples successively introduced into the analyser is avoided. Two general groups have been described by Valcarcel and Luque de Castro [20]. These are illustrated in Table 2.4.

Table 2.4 Classification of continuous automatic methods (after Valcarcel and Luque de Castro [20])

According to whether or not the streams contain bubbles	According to procedure	Sample	Introduction/ Nature of flow	Name
Segmented	By aspiration	Sequential	Continuous	Segmented flow analysis (SFA)
	By injection	Sequential	Continuous	Flow-injection analysis (FIA)
		Sequential	Discontinuous	Stopped-flow kinetic methods
Unsegmented	By aspiration	Continuous	Continuous	Completely continuous-flow analysis (CCFA)
		Sequential	Continuous	
		Sequential	Discontinuous	Controlled-dispersion flow analysis (CDFA)

Continuous segmented methods avoid carry-over by use of air bubbles establishing physical separations (segments) along the continuous flowing stream. These methods were invented by Skeggs [1] and formed the basis of the Technicon AutoAnalyzer. They are now also implemented on Skalar assemblies. Samples are introduced sequentially by aspiration with a moving articulated pipette.

Continuous unsegmented methods are characterized by the absence of air bubbles from the flowing system and by their greater technical simplicity. The way in which carry-over is avoided differs from mode to mode. In flow-injection analysis [21,22], the samples are introduced sequentially by injection or insertion of a preset volume into an uninterrupted liquid stream of reagent or carrier-diluent. In stopped-flow kinetic methods, the samples are injected simultaneously with the reagent and the mixtures of both are transported at high speed to the measuring cell, where the flow is stopped for as long as detection lasts.

The entire analytical operation - including data collection - thus takes place in a very short time (of the order of a few hundredths or thousandths of a second) [23]. These methods are used to study the kinetics of physico-chemical reactions using complex, precise and costly commercial configurations. In completely continuous methods, the sample and reagents are aspirated continuously into the flow system and no discontinuities resulting from injection, sample changeover or flow halting are introduced [24]. In some continuous unsegmented methods with sequential continuous aspiration of the sample, this is introduced until the transient signal sought is obtained, the next sample is introduced to provide a discontinuity, distinguishing these from completely continuous methods. The final category is termed 'controlled-dispersion flow methods'. In these methods intermittent pumping serves the same function as the injection valve does in continuous segmented methods [25].

2.5.1 Technicon AutoAnalyzer continuous-flow analysers

Automatic air-segmented analysers are characterized by the use of one or several liquid streams (diluents, washing solutions, reagents) where the sequentially aspirated samples are introduced and spaced by means of air bubbles aimed at avoiding the undesirable carry-over [26-28].

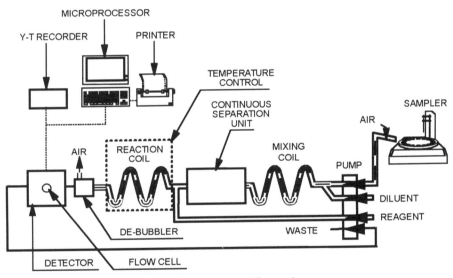

Fig. 2.13   Automatic continuous segmented flow analyser.

This type of analyser is usually modular in nature and consists of a series of elements (apparatus, instruments) coupled on-line to one another. Figure 2.13 shows the essential components of a straightforward Technicon AutoAnalyzer, namely:

Sampling system, which consists of a sample turntable and moving articulated aspiration probe, and differs very little from the generic designs.

Propelling unit(s), aimed to 'move' the samples. They are generally peristaltic pumps, although their function can also be served by piston pumps and the pressure exerted by a gas or gravitational force. They are intended to set and keep several streams in motion - the flow-rate of such streams should be regulatable and maintained as constant as possible (normally accomplished by using flexible tubes that withstand the mechanical pressure to which they are subjected).

Reaction-mixing coils, pieces of polyethylene, PTFE or glass tubing where the mixing of reactants and the analytical reaction take place. Their length and the flow-rate of streams circulated through them determine the time over which the reacting mixture 'resides' in them and hence the sampling frequency, inversely proportional to the time required for the complete analytical reaction.

Heating system, usually consisting of thermostatted baths or electrical wires wrapping the coils to favour the development of the analytical reaction.

Continuous separation systems, optional elements such as dialysers, liquid-liquid extractors, sorption or ion-exchange columns and filters, that can be placed before the reaction coils to remove potentially interfering species.

Debubbler, which removes the previously introduced air bubbles in order to avoid parasitic signals being produced by their interfaces upon reaching the detector. They are not normally required in the more recent designs as the signals from the detector are usually handled by a computer capable of discriminating between these undesirable signals and those actually corresponding to the reaction mixture.

Continuous detection system, usually of optical (colorimetric, photometric, fluorimetric) or electroanalytical (potentiometric, voltammetric) nature. The design of the flow-cell, when required, must be suited to the particular detection system used.

System for data collection and treatment, which should be able to operate in a continuous fashion and be as simple as a chart recorder or as sophisticated as an advanced microprocessor carrying out both operations and eventually delivering the results as required.

Although some of these elements, such as the heater, continuous separator, debubbler or microprocessor, are not indispensable, the rest are fundamental to the design of continuous segmented analysers. Although developed primarily for a clinical market, the concepts found many applications in the industrial area as well. The Laboratory of the Government Chemist (LGC) has established many automatic analyses for routine use using these concepts [29-31].

The major features of a determination carried out on an automatic segmented-flow analyser, namely precision and rapidity, are highly influenced by technical factors such as the extent of carry-over and mixing of reactants, and the time during which the reacting plug remains in the system.

The performance of an analyser which processes discrete samples at intervals is related to the dynamics of the flowing stream, and an understanding of the dominant factors is important in optimizing the design of continuous methods. A continuous stream of liquid flowing through a tube exhibits a velocity profile, the flow being fastest at the centre and slowest at the tube surface where frictional retardation occurs. Material at the periphery mixes with that in the centre of the following liquid and is the cause of sample carry-over, referred to briefly above. Segmentation of the stream by air-bubbles reduces carry-over by providing a barrier to mixing, but it does not entirely prevent it, because mixing in the surface layer can still occur. Nevertheless, carry-over occurs mainly in unsegmented streams and in terms of the standard AutoAnalyzer design this implies the initial sample-line before air-segmentation and after de-bubbling before entering the detector. The need for quantitative study of the magnitude of carry-over as a function of the kinetic parameters of the analyser has prompted definitive studies by several groups [32-35].

Two parameters have been demonstrated to be fundamental in calculating the performance characteristics of a continuous analyser, the lag phase and the half-wash time; they afford a correlation between the approach to steady state, fraction of steady state reached in a given time and the interaction between samples. The half-wash time ($W_{1/2}$) is the time for the detector response to change from any value to half that value, the lag phase L is defined in the ensuing discussion.

The standard detector-response curve for a continuous analyser is shown in Fig. 2.14; it is obtained by aspirating a blank solution until a steady baseline is obtained, adjusting the baseline to read zero at the detector, introducing the sample for a fixed period and finally aspirating the blank again. If the sampling time is made sufficiently long, a steady-state plateau is reached; if not, a peak is reached at a fraction of the steady-state reading, the fractional value being a function of the sampling time. The detector response comprises a rise-curve, a steady-state plateau (which may or may not be obtained) and a fall-curve.

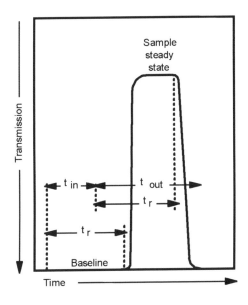

Fig. 2.14 Standard response curve for continuous flow analysis of discrete samples.

Detailed studies reveal that apart from an initial large phase, the rise-curve is exponential. Thus the measured concentration C as a function of time, t, is given by Eq. 2.1, where C_{ss} and C_t are the concentrations at the steady state and at time t.

Eq. 2.1
$$\frac{dC}{dt} = k(C_{ss} - C_t)$$

A plot of log C_t against time takes the generic form of Fig. 2.15. The value of $W_{1/2}$ is calculable directly from the slope of the linear portion of the plot. The initial non-exponential part of the plot is termed the lag phase L and is expressed numerically as the value of the intercept of the linear portion on the time axis. The fall-curve structure is the inverse of that of the rise-curve.

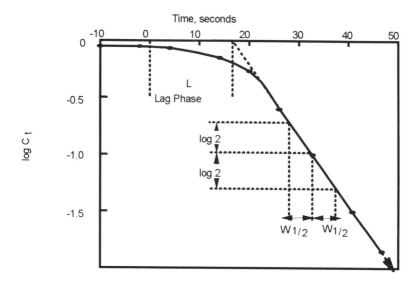

Fig. 2.15　　Graphical representation of lag phase L and half-wash time $W_{1/2}$.

In the continuous processing of discrete samples in the AutoAnalyzer system, the reaction-time is held constant by the manifold design, and because the rise-curve is exponential the degree of attainment of steady-state conditions is independent of concentration. Consequently it is unnecessary for the analytical reaction to proceed to completion for Beer's Law to be obeyed. This confers a considerable advantage upon the AutoAnalyzer approach and one which is frequently emphasized. The relationship between degree of attainment of steady state and $W_{1/2}$ can be generalized in the semi-logarithmic plot of Fig. 2.16 [10], where time is expressed in units of $W_{1/2}$.

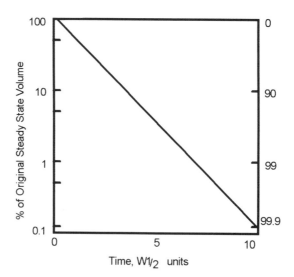

Fig. 2.16　　Influence of time, expressed in units of $W_{1/2}$ on extent of achievement of steady-state conditions.

The exponential nature of the fall-curve is responsible for sample interaction. Provided that the between-sample time is made long enough, or blanks are inserted between samples, the effect is negligible, but if the sampling rate is increased then the response to any given sample will be influenced by the tail of the response of the preceding one. The effect can be particularly severe when a concentrated sample precedes a dilute one; the peak due to the latter may appear as a shoulder on the tail on the one due to the former, or, in severe cases, it may be entirely hidden.

Sample interaction can be quantitatively expressed by using $W_{1/2}$ and L. If the between-sample time is t_b, the value of the expression $(t_b-L)/W_{1/2}$, gives a measure of the interaction of a sample with the following one. For values of $(t_b-L)/W_{1/2}$ of 1, 2, 3, 4, - the degree of interaction with the following sample is 50, 25, 12.5, 10% and this interaction appears additively in the response for the following sample.

Clearly the smaller the values of L and $W_{1/2}$ the better is the performance of the analyser. For a fixed sampling rate, the lower the values of L and $W_{1/2}$ the lower is the degree of interaction, or conversely, for a given acceptable per cent interaction, the lower the values of L and $W_{1/2}$, the faster is the allowable sampling rate. The numerical relationship between sampling frequency, $W_{1/2}$ and degree of interaction for the case where L is negligible is plotted in Fig. 2.17 [11]. By using AutoAnalyzer methods for sugar, urea and isocitrate dehydrogenase, Walker, Pennock and McGowan [35] determined values of L and $W_{1/2}$ as a function of the apparatus components. Significant increases in $W_{1/2}$ were observed when the flow-rate through the colorimeter cell was reduced and when a debubbler with recycling was introduced. In the majority of cases, measurable sample-interaction is limited to a sample and the one immediately following it.

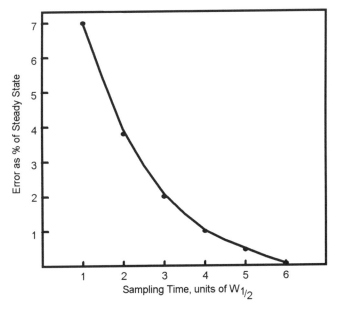

Fig. 2.17   Relationship between sampling time, $W_{1/2}$ and percentage interaction.

Wallace [34] demonstrated that a linear absorbance-to-concentration relationship is adequate to correct for the interaction. The absorbance A_n for sample n can be related to its true concentration by Eq. 2.2

Eq. 2.2
$$A_n = (1 - m)aC_n + mA_{n-1} + b$$

where a is the absorptivity and m the fraction of sample $n - 1$ in sample n, and b the blank absorbance.

2.5.1.1 Drift in continuous-flow analysers

A problem almost universally encountered in continuous-flow systems is that the instrument response for a given sample assay-value tends to vary with time. This effect, known as drift, affects the accuracy of results. It may be due to several causes, in particular variable performance of analyser components and variations in chemical sensitivity of the method used. It is manifest in two forms, baseline drift and peak-reading drift, which is due to sensitivity changes. The baseline drift may be detected visually if a chart record is kept, provided that the magnitude of the effect is sufficient. The peak-reading drift can only be detected if calibration standards are inserted at regular intervals between samples. Experience has shown that there are very few methods in which the effect of drift is negligible over a long period of operation and therefore the inclusion of periodic calibration standards is mandatory if errors due to drift are to be corrected. Indeed, in AutoAnalyzer usage it has become common to make every tenth reading a standard, and to calculate sample results with reference to the new standard response. This approach assumes that the drift is wholly systematic, and cannot take random fluctuations in the response to the individual standard into account. A more meaningful approach is to recalibrate with several different standards from time to time. Computer techniques are used to update the calibration graph to compensate for drift. Significant changes in instrument design and software algorithms have greatly improved the reliability of calibration graphs and data analyses.

Continuous-flow analysers have been extensively used, primarily in the clinical field but also in many industrial laboratory situations. They have also found use in process analysers. Newer generations of modules have focused attention on automated data processing and hands-off analyses, and automated shut down procedures, to exploit the use in silent hours. The extent of applications encompasses many simple monitoring systems and can also extend to fairly complicated systems, for example the system described by Burns [36] for analysing vitamin tablets. This includes a tablet dissolution system, an extraction device and an evaporation to dryness module, and the analysis is completed by coupling to an HPLC system. However, the sophistication and complication of this device has seemingly mitigated against its commercial acceptance. Basically, no two manufacturers' tablets need exactly the same treatment, and the system contravenes the first law of automation, i.e. keep it simple.

2.5.1.2 Developments in continuous-flow analysis

Skeggs' innovative step, the introduction of air bubbles into the flowing stream, attempted to minimize the time taken for a steady-state condition to be reached in the detector. The definitive description of dispersion in segmented streams (Snyder [37]) showed a complex relationship between internal diameter, liquid flow rate, segmentation frequency, residence time in the flow system, viscosity of the liquid and surface tension.

The most important conclusions from the proposed model are that (a) for any given tubing diameter there are optimum values for flow rate and segmentation frequency to achieve minimum dispersion and (b) dispersion can be decreased without limits by reducing tubing diameter.

Other sources of dispersion are the transmission tubing carrying the sample from the autosampler to the pump and the flow cell, where a debubbler removes the air segmentation before the liquid stream passes through the light path. The introduction of electronic or software mechanisms to eliminate interference from air bubbles passing through the flow cell eliminated the need for flow cell debubblers and was a major contributor to reducing total system dispersion. This is often known as 'bubble gating'. Sample line dispersion can be reduced by introducing several air bubbles between samples and the inter-sample wash, using a so-called pecking-probe sampler. In practice, this is most useful for viscous samples such as blood serum.

The various advances in system designed introduced by the knowledge of the above principles, have halved the dispersion and almost doubled the analysis rate for each new generation. For example, a typical method which uses dialysis and a heated reaction stage would run at 30 samples per hour on AAI systems, 60 samples per hour on AAII and similar systems and 120 samples per hour on third generation systems such as the Technicon SMAC, Alpkem RFA300 and the Bran & Luebbe TRAACS 800.

Other system improvements have centred on easier operation and higher automation. Computers, originally used only for calculating results, have taken over system control to automate functions such as colorimeter baseline and gain setting and the control of random-access samplers. These samplers allow samples which fall off-scale to be diluted and re-analysed in an extension to the original analysis run, thus avoiding the need for manual dilution and re-analysis. An accessory to the TRAACS system allows the reagents for several different tests on a multi-method manifold to be switched automatically between runs, allowing a single channel to be used for multiple analyses.

The TRAACS 800+ shown in Fig. 2.18 consists of one or more analytical consoles, a random-access sampler and personal computer (PC). The analytical console is fitted with two peristaltic pumps which can each handle up to 10 pump tubes. The analytical manifolds are located above the pumps and utilize glass parts 1 mm in diameter, in order to keep reagent consumption low. Heating baths, UV digestors and dialysers can be built into the manifold.

A continuously operating peristaltic pump delivers samples, standard solutions and reagents to the analytical system. A colour reaction then takes place in a heating bath or time delay coil, allowing reactions up to 10 minutes to go to completion.

Even complex procedures can be automated, such as: dialysis to clean up dirty samplers, solvent extraction, automatic distillation and on-line UV digestion. Unlike the earlier AutoAnalyzer systems which use a purely step-wise autosampler, the TRAACS is fitted with a random-access sampler as standard.

The random-access sampler can go to any sample cup position, any number of times, at any time during a run. This ability to sample cups in any order and to return to sample cups more than once, allows system automation to be greatly extended. It saves time and work by allowing automatic re-run of sample(s) following off-scale peaks and also the automatic dilution and re-analysis of off-scale samples. The sampler also saves cup positions, allowing more samples and longer unattended runs. For example, one set of standards provides initial calibration, drift correction, carry-over correction and periodic quality control. In addition, samples or standards can be sampled in replicate form from a single cup. The random-access sampler can be controlled and either the operator or the computer can make the decision as to which cup the sampler must go to.

A further benefit in everyday operation is that samples and standards can be separated on the sample tray, thus simplifying sample preparation and tray loading. The instructions for cup positions are simply entered with the other run parameters when the operator specifies the analysis conditions.

Multi-test cartridges are available for many methods. These allow several methods to be run on the same manifold. They are available for many types of sample, including seawater, boiler water, drinking water and waste water.

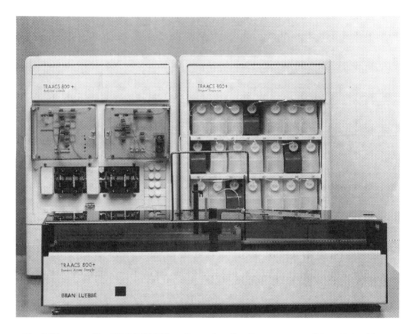

Fig. 2.18   The TRAACS 800+. *Reproduced with permission of Bran & Luebbe.*

The photometers are installed in the upper part of the analytical console. Each analytical channel is fitted with a double-beam filter photometer. The wavelength range is 340 - 900 nm. Flow cells are available in lengths of 10, 30 and 50 mm. Detectors such as a flame photometer, conductivity cell or pH electrode can also be connected externally when required.

The TRAACS 800+ is controlled by a personal computer and the features provided include: complete interactive control via keyboard or mouse; calculation of results as necessary taking into account baseline or sensitivity drift, graphical output of calibration curves for all calibration types - either linear or non-linear, input facility for sample identification data; allowing storage on disc and real-time results together with chart traces on a computer printer. The programs allow easy access to input or data files and connection to other computers, and gives system performance verification to CLP standards and built-in QC charts.

On completion of the analysis the operator or computer can decide that a particular cup should be re-sampled. For example this can occur when a very high concentration sample has gone off-scale: the sample or samples following the off-scale peak may be affected by carry-over resulting in a falsely high value. In this case the computer can automatically instruct the system to return to the cups following the off-scale samples and to re-sample them. A further possibility exists on systems fitted with Automatic Dilution. In this case, after the re-sampling process described has taken place, the off-scale samples themselves, together with the necessary high standards, are re-sampled through the dilution system. The results for all of the re-run and diluted samples are incorporated into the final report.

The system described above has been designed to increase laboratory productivity. It is claimed to have 30 - 70% lower reagent consumption than macroflow or flow injection systems. It also provides reagent change-over facilities, wash-out and shut-down sequencing, when the option is fitted.

2.5.2 Flow-injection analysers

These analytical systems pioneered by Stewart *et al.* [38,39] and Ruzicka and Hansen [40] superficially fulfil the aim of simplicity.

The considerable number of publications on FIA are evidence of its rapid growth and wide applicability. Since the initial work in 1975, the developments have been spectacular, leading to more than 600 publications, several excellent reviews and two monographs. The number of working routine systems is disappointingly low, owing to several problem areas which are discussed below.

Basically, FIA is a type of continuous-flow analysis in which the flow is not segmented with air bubbles. A quantity of dissolved sample, accurately dispensed, is injected or introduced into the carrier stream flowing through the FIA system. Chemical reactions or extractions which occur between the sample and the carrier can be added. As the analyte passes through the continuous detector, a transient signal is generated and recorded. The measurement is carried out under non-equilibrium conditions, since neither equilibrium (steady-state) nor homogeneity are realized prior to detection. This provides an advantage for FIA in that the response does not have to be any more than reproducible. This allows it to be used to measure trends, especially for process analysis. Stewart and Rosenfeld [39] have described a simple approach to measure pH. This approach gives a measurement of pH as a simple timing measurement. The signal deviates from the baseline as the sample emerges through the detector. The pH is characterized by the time the signal remains above the baseline level. Measuring this value is a simple function for modern microcomputer technology to quantify.

FIA is a fixed-time analytical methodology, since neither physical equilibrium (homogenization of a portion of the flow) nor chemical equilibrium (reaction completeness) has been attained by the time the signal is detected. The operational timing must be highly reproducible, because the measurements are made under non-steady-state conditions, so that small changes may give rise to serious errors in the results obtained.

A simplistic view of a FIA analytical system is outlined in Fig. 2.19. The liquid flow can be obtained in a number of ways, most commonly by using a peristaltic pump. In addition, gravity-feed systems, overpressure systems on liquid vessels and simple and double piston pumps normally associated with HPLC systems, are often used. The interrelation of the pumping system and the bore size of the transport tubing to a great extent modify the theoretical considerations involved and the operation of the FIA regime. In common with CFA, the minimization of dead volumes - in detector cells and between T-pieces for example - is particularly important. Injections of the sample into the flow line are usually accomplished by using six-port rotary inert valves. For most

FIA systems, the reaction is carried out in the normal tubing diameter, although chambers or coil reactors have been mixed. Their use, however, often obviates the principles of FIA. The primary requirement is to use a small-volume flow cell which is swept clean with minimal dead volume to provide a usable transient signal. The transient signals are characterized by the peak height or area residence or response time, or peak width at half height or at the baseline.

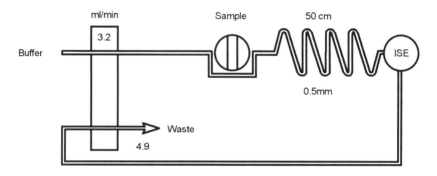

Fig. 2.19   Schematic diagram of a simple flow-injection system.

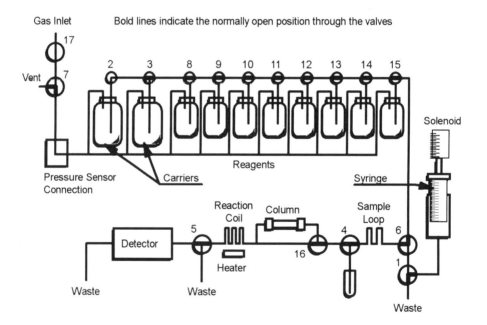

Fig. 2.20   Schematic diagram of simple flow-injection system designed by Malcolme-Lawes [41].

When peak-width at the baseline is used, the overall shape of the peak is insignificant and the simple measurement of time characterizes the analytical signals. Originally, two FIA approaches were introduced - the American AMFIA System which attempted to provide an automated system using HPLC instrumentation, and a manual unit which was developed by Ruzicka and Hansen [40]. A novel development of FIA has recently been demonstrated by Malcolme-Lawes and Pasquini [41] and a schematic diagram of this is shown in Fig. 2.20.

Figure 2.21 shows a basic FIA configuration consisting of: (1) a pumping unit, which drives one or several flowing solutions (either containing a dissolved analyte or just acting as carriers) at a constant rate; (2) an injection system which allows the reproducible insertion or introduction of an accurately measured sample volume into the flow without halting it; (3) a piece of tubing, usually referred to as the reactor, along which the reacting mixture is transported; (4) a flow-cell accommodated in a measuring device (photometer, fluorimeter, potentiometer) which transduces the signal and sends it to a recorder and/or microprocessor. Usually, the stream emerging from the sensing system is directed to waste.

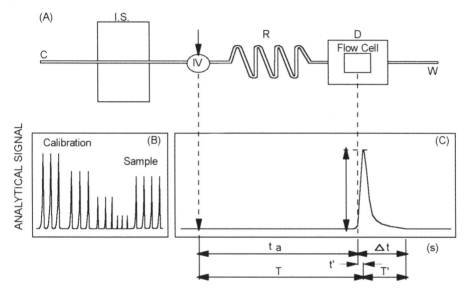

Fig. 2.21 (a) Scheme of a basic FIA system. C: carrier (with or without a dissolved reagent); PS: propelling system; IV: injection valve; R: reactor; D: detector; W: waste. (b) Slow recording - the commonest in this technique - corresponding to injection of triplicate standards and quintuplicate samples. (c) Fast recording, with its characteristic parameters.

Figure 2.21 also shows typical recordings obtained with this technique.

The following parameters are obtained:

1. The peak height, which is directly related to the analyte concentration.

2. The Residence time, T, defined as the interval elapsed from injection to the appearance of the maximum signal.

3. The Travel time, t_a, the period separating injection from the end of the baseline.

4. The Return time, T', the time elapsed between appearance of the maximum signal and baseline restoration.

5. Baseline-to-baseline time, Δt, the interval elapsed between the onset of the peak and baseline restoration.

FIA peaks are not Gaussian curves, so that the parameters above do not describe the peak shape in full - particularly the trailing portion, which is peculiar to this type of recording and distinguishes it from the transient signals typical of other analytical techniques.

Flow-injection and continuous-flow systems are very similar. The major differences are outlined here. Continuous-flow systems are characterized by a relatively long start-up time prior to instrument stabilization, whereas the flow-injection approach requires little more time than that needed to stabilize the detector output. Tubing diameters on a flow-injection manifold are usually much smaller and the samples are injected into the flow line rather than aspirated. No wash cycle is employed in the flow-injection regime, since the sample is a discrete slug. Flow rates in continuous-flow manifolds are often larger than in the flow-injection regime.

Whilst similar detection systems are used for both techniques there is a fundamental difference as the detection takes place. Continuous-flow uses steady-state conditions, whereas the FIA measurements are made in non-steady-state conditions. A major advantage of the FIA regime, an area hitherto not fully exploited, is the time-frame for an analytical measurement. That is to say the result is available by FIA much faster than in the continuous regime. Therefore the major areas of interest should undoubtedly be for process analysis where trends can be readily observed on a rapid basis.

Valcarcel and Luque de Castro [20] have made a comparison of the parameters of the two techniques, and these are set out in Table 2.3.

Table 2.3 Comparison between the two chief types of continuous analytical methodologies: segmented (SFA) and unsegmented (FIA)

Feature	SFA	FIA
Sample introduction	Aspiration	Injection
Sample volume	0.2 - 2 ml	10 - 100 µl
Response time	2 - 30 min	3 - 60 s
Tubing inner diameter	2 mm	0.5 - 0.7 mm
Detection	In equilibrium (homogeneity)	With controlled partial dispersion
Sample throughput	≤80 samples/hr	≤300 samples/hr
Precision	1 - 2%	1 - 2%
Reagent consumption	High	Low
Washout cycle	Essential	None
Continuous kinetic analysis	Not feasible	Stopped-flow
Titrations	No	Yes
Data supplied	Peak height	Peak height, area and width; peak-to-peak distance

2.5.2.1 Theoretical background to flow-injection analysis

The physical foundation of FIA is related to the behaviour of the sample plug inserted in the flow, which is characterized mathematically by means of **dispersion**. This, in turn, is defined by the shape of the profile yielded by the injected sample portion along the system, particularly at the flow cell.

Flow-injection was first thought to be operating in the turbulent regime, but recently it has been demonstrated beyond doubt that most FIA systems operate under laminar flow conditions [42,43].

When the sample is injected into the flowing stream, its dilution within it is a function, of, among other factors, the flow rate and geometry of the transport system. Initially, dilution is chiefly convective in nature; later it becomes convective-diffusional and eventually develops a purely diffusional character. The intermediate situation (i.e. convective-diffusional transport) is by far the most common in FIA.

It has been demonstrated that radial dispersion contributes more significantly to the dilution of the sample in the flow than does axial dispersion. This type of fluid movement, termed 'secondary flow' by Tijssen [43], results in a washout effect accounting for the low mutual contamination of samples successively injected into a carrier stream. This advantageous feature is a result of the use of low flow rates and small tubing bores, and results in decreased peak-width and hence to increased sampling rate.

Theoretical studies on FIA have been aimed at establishing an accurate relationship between the geometric (length and diameter) and hydrodynamic (flow rate) characteristics of a FIA system and the parameters defining the profile obtained (travel time, co-ordinates of the peak maximum and peak width).

CONVECTIVE TRANSPORT

DIFFUSIONAL TRANSPORT

Fig. 2.22   General types of transport in closed tubes.

Several models have been proposed to account for the non-Gaussian shape of typical FIA recordings, which reflects the behaviour of a solute injected into a FIA system. The so-called *General model* is the best way of describing the behaviour of a solute injected into a FIA system. It is based on the general expression describing convective-diffusional transport, which takes account of both axial and radial concentration gradients, the parabolic shape of the velocity profile corresponding to a laminar flow regime and the contribution of convective transport:

Eq. 2.3

$$D\left[\frac{\delta^2 C}{\delta x^2} + \frac{\delta^2 C}{\delta r^2} + \frac{1}{r}\frac{\delta C}{\delta r}\right] = \frac{\delta C}{\delta t} + Uo\left[1 - \frac{r^2}{a^2}\right]\frac{\delta C}{\delta x}$$

where D is the molecular diffusion coefficient (expressed in cm^2/s), C is the analyte concentration, t is the time (in s), x is the distance from the tube radius (cm) and U_o is the linear flow velocity (cm/s) [44].

This equation is by no means easy to solve in a straightforward manner. By use of suitable approximations, Vanderslice et al. [42] derived two expressions relating two of the most important parameters of the FIA curve (travel time and peak width) to the essential characteristics of an ordinary FIA system (namely the reactor length, l, and radius, r, and the flow-rate, q), together with the molecular diffusion coefficient:

$$t_a = 109 r^{20.025}(L/q)^{1.025}(1/f)$$

$$\Delta t = 35.4(r^2/D^{0.36})(L/q)^{0.64} f$$

These two expressions are descriptive of experimental facts only if an accommodation factor, f, is included.

A somewhat different approach to the problem was derived by Ruzicka and Hansen [22]. More recently, Panton and Mottola [44,45] have studied the chemical contribution to dispersion.

2.6 CONCLUSIONS

The philosophies for automation have been described in the foregoing sections. However, to solve an analytical problem there may well be more than one approach that offers potential. The literature abounds with methods that have been automated by flow-injection and by continuous-flow methodologies. Also, very often a procedure which involves several stages prior to the actual measurement can be configured by combining two of the approaches. An example of this is the automated Quinizarium system described by Tucker et al. [46]. This was a continuous extraction followed by a batch extraction which is finally completed by a batch measurement on a discrete sample for quantification and measurement. Whereas sample preparation is almost always required, there is no doubt in my mind that the best approach to this area of activity is to avoid it totally. The application of near infra-red spectroscopy is an example of this strategy.

Applications of near infra-red reflectance spectroscopy were first introduced by Norris and Hart [47] for the determination of moisture, oil and fat in cereal products. Several instrument companies, notably Bran & Luebbe, formerly Technicon, have developed commercial instruments. The instrument replaces a series of chemical procedures by signal measurement in each of six infra-red regions, and reference to a suitable computer calibration. Such an approach offers considerable advantages and is novel. It is made more attractive by the availability of microprocessor computing power. Although the instrumentation has been specifically developed for the cereal market, it has much wider applications, for example in the tobacco industry. It does, however, suffer from some disadvantages. The instrument must be calibrated against a suitably accurate standard method, and only a few chemical methods are available. Also, the six wavelengths were selected for the prime objective of wheat analysis, and are not the most appropriate for other types of analyses. On the other hand, one of the major advantages of the technique is that the analysis can be done on site, away from the laboratory.

Evaluation and use of the first generation of reflectance infra-red instruments has indicated the specification necessary for an instrument with more flexibility.

The requirements are as follows:

1. Fast, simple, flexible sample-handling.
2. Simple to operate, standardize and calibrate.
3. Capability to analyse many different products and for many constituents.
4. Ability to operate safely and accurately over wide-ranging environmental conditions.
5. High accuracy in detecting energy reflected from the sample.
6. Insensitivity to variable sample matrix effects such as particle-size variations.
7. Measurement at several specific wavelengths with narrow bandwidth to remove interferences and provide highest accuracy for the constituent measured.
8. Ability to adapt to new advances in the technology of optical measurement, signal processing and data treatment.
9. Capability for rapid, easy diagnosis and correction of instrument malfunctions.

The InfraAnalyser 400 has been designed to meet these requirements and is a good example of how, by the introduction of current technology, a flexible instrument results. In addition, an integrating sphere detection system has been incorporated into the design, which, coupled with Kohler optics (as used in microscopy), allows uniform sample illumination and good signal-to-noise characteristics. The new instrument is also more independent of sample particle-size and temperature. An integral sealed gold standard is fitted, which ensures minimal drift and allows absolute readings of reflectance data to be made. Up to 20 different wavelength detectors with narrow bandwidths are available, including one for diagnostic purposes. The incorporation of microprocessor technology allows such facilities as self-calibration and the introduction of self-teaching aids. The computer system is also fully expandable to cater for a range of applications. Applications in wheat, dairy, animal feed, tobacco, and cocoa analysis and many other areas are under evaluation. Certainly some of the major obstacles to the application of the technique to a wider range of samples, especially the inflexibility, have been overcome by this instrument and subsequent models.

REFERENCES

[1] Skeggs, L.T., *American Journal of Clinical Pathology*, 1957, 28, 311.
[2] Betteridge, D., Dagless, E.L., David, P., Deans, D.R., Penketh, G.E. and Shawcross, P., *Analyst*, 1976, 101, 409.
[3] Malmstadt, H.V., Enke, C.G., Crouch, S.R. and Horlich, G., *Electronic Measurements for Scientists*, Benjamin, Massachusetts, USA, 1973.
[4] Pittsburgh Conferences on Analytical Chemistry and Applied Spectroscopy. The Pittsburgh Conference, Suite 332, 300 Penn Center Boulevard, Pittsburgh, Pennsylvania 15235-5503, USA.
[5] Betteridge, D.M, Porter, D.G. and Stockwell, P.B., *Journal of Automatic Chemistry*, 1979, 1, 69.
[6] Porter D.G. and Stockwell, P.B. in *Topics in Automatic Chemical Analysis*, Edited by Stockwell, P.B. and Foreman, J.K., Horwood, Chichester, 1979.
[7] Carlyle, J.E., *Journal of Automatic Chemistry*, 1979, 1, 69.
[8] *Journal of Automatic Chemistry*, founded 1979, published by Taylor & Francis Ltd, 4 John Street, London WC1N 2ET.
[9] Arndt, R.W. and Werder, R.D., *Z. Analytical Chemistry*, 1977, 287, 15.
[10] Curie, H.G., Columbus, R.L., Dapper, G.M., Eder, T.W., Fellows, W.D., Figueras, J., Glover, C.P., Goffe, C.A., Hill, D.E., Lawton, W.H., Muka, E.J., Pinney, J.E., Rand, R.N., Sonford, K.J. and Wu, T.W., *Clinical Chemistry*, 1978, 24, 1335.
[11] Spayd, R.W., Bruschi, B., Burdict, B.A., Dapper, G.M., Eikenberry, J.N., Esders, T.W., Figueras, J., Goodhue, C.T., La Rossa, D.D., Nelson, R.W., Rand, R.N. and Wu, T.W., *Clinical Chemistry*, 1978, 24, 1343.
[12] Anderson, N.G., *American Journal of Clinical Pathology*, 1970, 53, 778.
[13] Burtis, C.A., Johnson, N.F., Attrill, J.E., Scott, C.D., Cho, N. and Anderson, N.G., *Clinical Chemistry*, 1971, 17, 686.
[14] Jansen, J.M., *Clinical Chemistry*, 1970, 16, 515.
[15] Kelley, M.T. and Jansen, J.M., *Clinical Chemistry*, 1971, 17, 701.
[16] Tiffany, T.O., Jansen, J.M., Burtis, C.A., Overton, J.B. and Scott, C.D., *Clinical Chemistry*, 1972, 18, 829.
[17] Snook, M.E., Renshaw, A., Rideout, J.M., Wright, D.J., Baker, J. and Dickins, J., *Journal of Automatic Chemistry*, 1979, 1, 72.
[18] Pardue, H.L., *Clinical Chemistry*, 1977, 23, 2189.
[19] Haeckel, R., Sonntag, O. and Petry, K., *Journal of Automatic Chemistry*, 1979, 1, 273.
[20] Valcarcel, M. and Luque de Castro, M.D., *Automatic Methods of Analysis*, Elsevier, Amsterdam, 1988.
[21] Valcarcel, M. and Luque de Castro, M.D., *Flow Injection Analysis - Principles and Applications*, Ellis Horwood, Chichester, UK, 1987.
[22] Ruzicka, J. and Hansen, E.H., *Flow Injection Analysis*, Wiley, New York, 1981.
[23] Perez-Bendito, D. and Silva, M., *Kinetic Methods in Analytical Chemistry*, Ellis Horwood, Chichester, UK, 1988.
[24] Goto, M., *Trends in Analytical Chemistry*, 1983, 2, 92.
[25] Rocks, B.F., Sherwood, R.A and Riley, C., *Analyst*, 1984, 109, 847.
[26] Furman, W.B., *Continuous Flow Analysis Theory in Practice*, Marcel Dekker, New York, 1976.

References

[27] Coakley, W.A., *Handbook of Automated Analysis*, Marcel Dekker, New York, 1981.
[28] Snyder, L., Levine, J., Stoy, R. and Conneta, A., *Analytical Chemistry*, 1976, 48, 942A.
[29] Sawyer, R. and Dixon, E.J., *Analyst*, 1968, 93, 669.
[30] Sawyer, R. and Dixon, E.J., *Analyst*, 1968, 93, 680.
[31] Sawyer, R., Dixon, E.J., Lidzey, R.G. and Stockwell, P.B., *Analyst*, 1970, 95, 957.
[32] Thiers, R.E. and Oglesby, K.M., *Clinical Chemistry*, 1964, 10, 246.
[33] Thiers, R.E., Cole, R.R. and Kirsch, W.J., *Clinical Chemistry*, 1967, 13, 951.
[34] Wallace, V., *Analytical Biochemistry*, 1967, 20, 517.
[35] Walker, W.H.C., Pennock, C.A. and McGowan, G.K., *Clinica Chimica Acta*, 1970, 27, 421.
[36] Burns, D.A, *Proceedings of the Seventh Technicon International Congress*, New York 1977.
[37] Snyder, L.R., *Journal of Chromatography*, 1976, 125, 287.
[38] Stewart, K.K., Beecher, G.R. and Hare, P.E., *Analytical Biochemistry*, 1976, 79, 162.
[39] Stewart, K.K, and Rosenfeld, A.G., *Analytical Chemistry*, 1982, 54, 2368.
[40] Ruzicka, J. and Hansen, E.H., *Analytica Chimica Acta*, 1975, 78, 31.
[41] Malcome-Lawes, D.J. and Pasquini, C., *Journal of Automatic Chemistry*, 1988, 10, 192.
[42] Vanderslice, J.T., Stewart, K.K., Rosenfeld, A.G. and Higgs, D.J., *Talanta*, 1981, 28, 11.
[43] Tijssen, A., *Analytica Chimica Acta*, 1980, 114, 71.
[44] Panton, C.C. and Mottola, H.A., *Analytica Chimica Acta*, 1983, 154, 1.
[45] Panton, C.C. and Mottola, H.A., *Analytica Chimica Acta*, 1984, 158, 67.
[46] Tucker, K.B.E., Sawyer, R. and Stockwell, P.B., *Analyst*, 1970, 95, 730.
[47] Norris, K.H. and Hart, J.R., *Proceedings of the 1963 International Symposium on Humidity and Moisture*, Reinhold, New York, 1965.

3

Total systems approach to laboratory automation

3.1 CASE STUDY I - REGULATION OF TAR AND NICOTINE IN CIGARETTES

3.1.1 Introduction

This chapter illustrates the use of the total systems approach to develop and implement complete analytical schemes. The first example shows how the Laboratory of the Government Chemist (LGC) was able to meet a requirement for twice-yearly 'league tables' of the tar and nicotine yields of cigarettes sold in the UK. The instrumentation described was installed in a period between 1972-1978, however, the problem involves many aspects of automation and serves to illustrate a number of useful points, especially relating to specification details. The development was stepwise, and it passed through an initial phase where classical manual methods were used in conjunction with a computer bureau, which, with programs developed at the LGC, processed the analytical data. The achievement of total systems automation involves automation of the analytical methods and the generation of an on-line data processing system utilizing an in-house Rank Xerox RX-530 computer. Administrative and technical constraints influenced the development philosophy, which is described in detail. It will be clear to the reader that the solutions outlined in this chapter may not necessarily use computing techniques available now. Despite this, the lessons of automation are useful to the system designer and clearly show the advantages of taking a total systems approach. In addition, the size and scope of the survey changed markedly throughout the period of development. Close co-operation between the analysts involved in the day-to-day work, the managers and the Laboratory Automation Group, meant that it was possible to cope with these changing demands. The constraints came from government and industry; a desire to have the manufacturers' agreements to the publication of the data required considerable additional consultancy at all stages. Whenever a new approach was to be used, it had to be fully validated before introduction into the scheme.

The requirement arose from the acceptance by the Government of the report *Smoking and Health Now* by the Royal College of Physicians [1]. The report concluded that the quantitative association between cigarette smoking and lung cancer is most simply explained on a causal basis. There is good presumptive evidence that the deposition of tar in smokers' lungs is an important factor in causing lung cancer, chronic bronchitis and emphysema, and the nicotine inhaled by cigarette smokers is a causative factor in cardiovascular diseases. The report recommended that the tar and nicotine content of the smoke of all brands of cigarettes should be published and that packets of cigarettes should be labelled with this information. The LGC was asked, in the early

1970s, to undertake the work of determining the tar and nicotine yields of cigarettes. This required carefully standardized and reproducible techniques if the results were to be of value for comparative purposes.

The terms 'tar' and 'nicotine' require definition. The smoke generated during the smoking of cigarettes and inhaled by the smoker (mainstream smoke) contains tarry matter, together with water and alkaloids from the nicotine group, of which nicotine is the major component. Tar is defined as the total particulate matter in the mainstream smoke. Both tar and nicotine are usually quantified as mg per cigarette when the cigarette is machine-smoked under standard conditions.

At first sight, the analysis of cigarette smoke appears to be a relatively simple chemical procedure. However, smoke analysis is fraught with problems which arise from the stepwise nature of the combustion process in which a volume of air is drawn through a column of inhomogeneously distributed tobacco constituents in short bursts (puffs), separated by relatively much longer periods of quiescent smouldering. The regime of heating to which the tobacco is subjected is a function of the airflow/pressure drop relationship generated by the smoking machine and is subject to modification by variations in the cigarette being tested and by the trap used to collect the material for subsequent chemical analysis. Standardization of all smoking variables is therefore an essential prerequisite of any laboratory procedure for cigarette smoke analysis. How better to achieve reproducibility of operation than by automation?

The methodology initially adopted by the LGC followed, for consistency, the manual techniques developed by the UK's tobacco industry [2]. Mechanization, as distinct from computerization, was confined initially to the use of a commercial 20-channel smoking machine (Filtrona Model 300) on which 20 separate brands of cigarette are smoked in a controlled, reproducible manner. This was because the tobacco industry used similar procedures, and newer automated methods had first to be validated. The mainstream smoke generated from each cigarette was drawn through a glass fibre filter pad and the particulate matter collected was analysed by labour-intensive manual techniques. Interaction with the skills of the automation group played a significant part even at this preliminary stage, and some of the equipment used was redesigned to improve its operational characteristics, and, hence, the reproducibility of the manual analyses.

The production of a single tar and nicotine table in the UK requires the smoking and analysis of some 20 000 cigarettes, and this generates a large amount of data. An essential prerequisite of this, or of any other large-scale routine analytical work, is a computer-based data processing facility, and from its inception the data collecting and reporting were designed around the LGC's computer facilities. The design of this first-stage data processing system was crucial to the subsequent development in overall automation of the analytical techniques. The total involvement of the analytical chemists, the systems designers and the programmers at all stages ensured a data processing facility that was flexible enough to serve the precise requirement of the initial manual analytical procedures, yet could be extended as the various aspects of automation were introduced. Although the computers used in the 1970s have become outdated, the overall design principles have not changed.

Data from manual laboratory techniques had to be transferred from handwritten laboratory work sheets to a computer file via an intermediate stage, such as punched cards or paper tape, and by the very nature of this form of batch processing the control functions were largely in the hands of the computer staff. However, as the analytical equipment becomes linked on-line to the computer, so control can be returned to the analyst. In the LGC, with the aid of very comprehensive interactive housekeeping programs, the analyst has fast, flexible and comprehensive control over the work program to a degree that could never have been achieved without the aid of automatic methods of analysis and computers. This applies particularly to database organization, instant recall of data with statistical assessments, reporting of results and monitoring performance of staff and equipment. The involvement of the computer gave the analysts more job satisfaction as well as giving the laboratory management more quality control.

3.1.2 Sampling

In the development of the protocol for the analysis of commercially manufactured cigarettes for comparative smoke analysis, a large number of variables must be taken into account if the results are to truly reflect not only any statistically significant difference between one brand and another, but also the yield for each brand throughout the sampling period. This latter point is particularly important when packets are printed with tar yield, or, in the case of the UK, with the tar group description taken from the latest government league table. To arrive at a satisfactory sampling system, it is necessary to have some understanding of the character of these in-built variables. It is important, as for all analytical procedures, that the sampling pattern gives meaningful results which do not distort the levels of tar and nicotine.

Tobacco is a product of the soil and the air, and is subject to all the variability in chemical composition associated with natural products. Its country of origin, the climate of the growing area, the genetic make-up of the variety of seed, the local agronomy and curing practice all play their part in determining the precise chemical composition of the tobacco purchased for cigarette manufacture. Expert purchasing and blending by the cigarette manufacturer, however, reduces these variables quite considerably.

The cigarette manufacturing process adds another variable to the final product through bulk blending and the use of cigarette papers, filters and tipping paper. Also, variations in the tobacco-rod to filter-rod ratio of each cigarette, and the packed density of the tobacco rod, are inherent in the cigarette making process, and they affect the weight of tobacco available for combustion and the draw and filtration characteristics of the cigarette (see Fig. 3.1). These variables have always been very tightly controlled by British manufacturers, particularly with respect to packing density and weight of tobacco, because tobacco duty was originally levied on the weight of tobacco in the cigarette. In most other countries the tobacco duty is charged on a 'per cigarette' basis, together with some *ad valorem* element, and there is therefore not the same incentive for the manufacturer to control at the same level. Many cigarettes imported into this country are more variable than their traditional home manufactured counterparts. However, in 1978 the tobacco duty system within the UK was harmonized with that of the European Union and is no longer based on the weight of the tobacco.

There are other important variables which are associated with the testing procedures in the laboratory. The human smoking regime is a highly individual and variable process, but this variability must be eliminated in the laboratory smoking regime without losing sight of the way cigarettes are smoked *in vivo*. The chemical procedures applied to the material collected from cigarette smoke introduce some further variability related to their own inherent accuracy and precision. However, there is little advantage in improving the accuracy and precision of these chemical procedures, or of introducing more precise methods, unless the variables associated with samples, sampling and smoking are built into the statistical design of the whole testing programme. The sampling procedure must be designed to produce cigarette samples for presentation to the smoking machine and analytical procedures that accurately reflect the actual composition and variation of each brand over the entire period of testing. This is difficult without a detailed knowledge of the domestic market position, and, in fact, the sampling procedure in the UK had to be radically revised in the light of experience gained during the first testing programme in 1972.

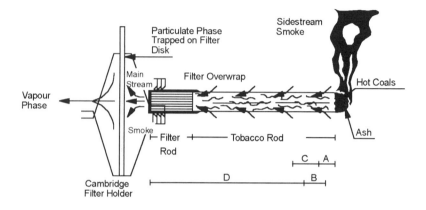

Fig. 3.1   Cigarette smoking regime.

By the design of the sampling procedure, the LGC, and the laboratories within the British tobacco industry, have reduced to an aceptable level the random variations associated with the raw material and manufacturing process. The variables associated with machine smoking and analysis have been largely eliminated by testing the samples in a random sequence and exercising close, and, where possible, automatic, control over all the smoking parameters and analytical procedures. Approximately 20 000 cigarettes are smoked and analysed to produce each league table. The constraints of sampling dictate that the analysis is organized over a period of six months, during which time 150 cigarettes are smoked from each of the major brands on sale. The cigarettes are smoked in groups of five taken from each packet of 20 sampled, and the total material collected from the smoke is analysed. Thus each league table comprises the mean value for tar and nicotine yields obtained from 30 separate packets of each brand.

The sampling procedure involves drawing a proportion of the total sample for each brand at regular intervals throughout the period of test. There are three possible locations: manufacturer's premises, wholesale warehouse and retail outlets. In 1972, samples for the first UK league table were drawn from retail shops because it was thought that this would most accurately reflect what the consumer was purchasing. The country was divided into five regions, and one packet of cigarettes of each brand was purchased at a high-turnover retail outlet within each region every month. Several large towns were visited each month, and these were varied from month to month. The results obtained from these samples revealed quite clearly that collecting samples from retail outlets in the UK was not a satisfactory method of sampling, primarily because of the very wide age range of the samples. As part of the testing programme the date of manufacture of all samples was noted, and this revealed a large number of very old and unrepresentative samples. These arose from a combination of poor stock rotation at some wholesale and retail outlets, and the national sales pattern of many brands. It is clearly unsatisfactory to allow results from samples obtained in areas of negligible sales to contribute an equal weighting to the mean smoke yields as samples obtained at high-turnover outlets: with an average manufacture-to-sale interval of no more than a few months, it is unrepresentative to include results from packets that are several years old. Indeed, samples of one brand spanned three years, and the analytical results showed no fewer than five distinct specifications, thereby distorting the final mean tar yield. So sampling at retail outlets was abandoned from 1973, and the second and subsequent league tables were based on five samples of every brand shipped by the manufacturer. This produces representative samples and ensures that the reported smoke yields more closely represent those packets available at retail outlets at the time the league table is published.

3.1.3 Manual laboratory procedures

Two categories of smoke are recognized when cigarettes are burned on a laboratory smoking machine. The smoke that emerges from the butt end of the cigarette during each puff is known as the 'mainstream' smoke and is this is what would normally be inhaled by the smoker. The smoke that is dispersed to the atmosphere during the smoulder period between puffs is known as the 'side-stream' smoke and is not included in this discussion.

Most laboratories involved in large-scale routine chemical analysis of mainstream smoke aim to produce results showing relative yields so that valid brand comparisons can be made. It is therefore essential to maintain careful standardization of all procedures, before and during laboratory smoking, in order to present comparably collected smoke to the subsequent analytical techniques. The quantity and composition of the mainstream smoke yielded by cigarettes is influenced by many potentially variable laboratory factors, the most important of which are discussed below and are standardized internationally [3,4].

3.1.3.1 Standard conditions

Changes in the moisture content of tobacco can affect the temperature of combustion within the hot coals and also the rate at which combustion occurs. Cigarettes must therefore be brought to equilibrium moisture content in a standard atmosphere.

The temperature and ambient relative humidity of the atmosphere in which the cigarettes are smoked influence the combustion process to a minor extent in a similar way to the influences of the moisture content of the tobacco. Ideally, the atmospheric conditions for laboratory smoking should be $22°C$ and 65% relative humidity. Air movement around the burning cigarettes will influence the combustion temperature and rate, particularly during the smoulder period, and should be adjusted so that the side-stream smoke is removed without increasing the smoulder rate of the cigarettes. This corresponds to a linear air flow of approximately 1 m/min at the cigarette.

The volume of air drawn through the burning zone during the application of a pressure difference to the butt end of the cigarette will have a substantial effect on the mainstream smoke yield. The laboratory smoking machine is required to take standard puffs for each cigarette.

The frequency with which the pressure difference is applied to the burning cigarette will affect the mainstream/side-stream smoke ratio and thus the total yield. The laboratory smoking machine must take a puff every 60 ± 1 s.

The length of time that the pressure difference is applied to the burning cigarette will, in a piston-operated smoking machine, affect the air flow-rate through the hot coals, and its effect is interrelated with puff volume and puff frequency.

The plot of air flow-rate as a function of time is called the puff profile. Variations in this will change the temperature profile within the burning zone and thus affect smoke yields. The laboratory smoking machine must reproducibly generate a bell-shaped puff profile to accepted standards.

The length of unburnt cigarette remaining after smoking is called the butt length, and it is the most important factor influencing smoke yield. Small differences in the point on the cigarette at which smoking on the machine ceases can give rise to very significant differences in yield. This can readily be explained in terms of the smoking process. As air is drawn through the burning zone during a puff, many of the products of volatilization, pyrolysis and recombination are filtered by, or condensed on, the strands of tobacco that they meet during the passage down the tobacco rod toward the butt end of the cigarette (see Fig. 3.1). The burning zone will encounter progressively more of the products from previous puffs, in addition to the basic tobacco matrix, as it moves down the cigarette; and this, coupled with the decreasing distance between the burning zone and the butt end, causes each puff to become successively richer in smoke products. It follows that the point of puff cessation will be critical, and variations of 1 - 2 mm will cause variations of 20% or more in the total tar yield. For reproducibility of operation, the smoking machine must therefore automatically standardize the puff termination point.

Puff termination on smoking machines is achieved by a cotton gate which lies across the cigarette at the butt mark, and also holds a microswitch in its active mode. On reaching the butt mark the burning zone breaks the cotton and releases the microswitch which in turn deactivates a solenoid valve that immediately prevents further pressure difference being applied to the cigarette. This is a simple, but effective, automatic solution to a difficult problem.

3.1.4 The smoking machine

The variables discussed so far can be reduced to an acceptable level by careful control and standardization, which, in the case of the smoking machine, is achieved largely by automatic operation. The remaining factors that influence smoke yields cannot be controlled in this way, and their effects have to be minimized by randomizing them across all samples. The samples tested during any six-month period are smoked in a random, but predetermined, order, so that day-to-day environmental and climatic changes, operator variability, and the effect of each of the 20 available smoking positions are largely cancelled out. This random smoking plan is produced by a computer.

Machine smoking at the LGC was carried out on a Filtrona 300 20-port Restricted Smoking Machine in which draughts, puff volume, puff frequency, puff duration, puff profile, and puff termination at the butt mark can be maintained reproducibly during the smoking sequence, provided that regular calibrations are carried out. The machine is also fully compensating, in that it maintains the preset smoking parameters over the range of pressure drop found in commercial cigarettes. It consists of 20 independent suction lines, all of which are connected to a common control and drive unit: each line has a discrete butt-termination device. Pressure difference is applied to the cigarette by the displacement of air within a piston connected to the drive unit by a crankshaft. Above each piston is a three-way solenoid valve that allows the negative pressure

Sec. 3.1] Regulation of tar and nicotine in cigarettes 73

created by the down-stroke of the piston to be applied to the cigarette (puff phase) but switches the positive pressure from the return stroke of the piston to a common manifold that discharges to atmosphere. Interposed between the cigarette and the solenoid valve is a Cambridge filter holder containing a disc of glass fibre, which removes all the particulate matter, greater than 0.3 μm diameter, from the mainstream smoke. The material passing through the Cambridge filter disc (referred to as 'vapour phase') is drawn into the piston and finally exhausted to atmosphere via the manifold. The mode of operation of these components of the smoking machine can be seen in the upper section of the diagram in Fig. 3.2 (a-d).

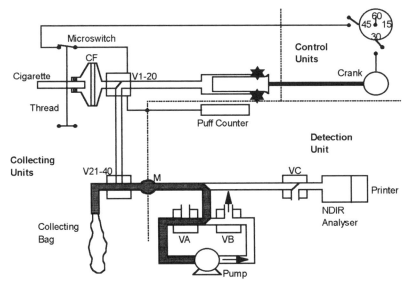

Fig. 3.2(a)   *Pre-smoking* - Cotton thread fixed to microswitch and aligned to butt mark on cigarette. Vapour-phase collecting bag flushed and evacuated.

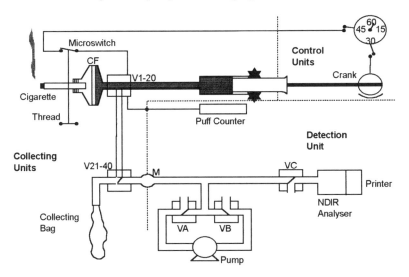

Fig. 3.2(b)   *Puff* - Crankshaft rotating, piston draws air through burning cigarette depositing particulate phase on Cambridge filter. Vapour-phase passes through to piston.

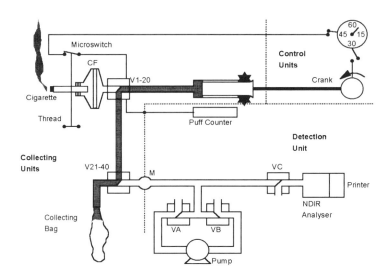

Fig. 3.2(c)   *Exhaust* - Puff ceased; piston exhausts vapour phase to collecting bag.

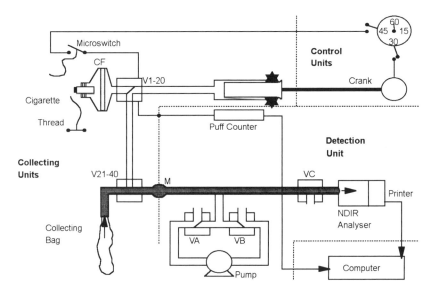

Fig. 3.2(d)   *Post smoking* - Cotton broken and microswitch released. Pump of NDIR Analyser samples vapour phase from collecting bag to give direct reading of carbon monoxide concentration. Puff counts and carbon monoxide concentration to the computer. Cambridge filter removed for chemical analysis.

KEY
CF   Cambridge filter
V1 - 20   20 primary 3-way solenoid valves (one per channel)
V21 - 40   20 secondary 3-way solenoid valves (one per channel)
VA, VB   3-way solenoid valves controlling the flush/evacuation cycle of collecting bags.
VC   3-way solenoid valve controlling access to NDIR Analyser
M   Manifold to which all 20 channels are connected.
NDIR   Nondispersive infra-red detector

Before a set of cigarettes is smoked, the displacement of each piston is adjusted. Five cigarettes from 20 separate samples, normally 20 separate brands, are then smoked consecutively according to the standard conditions described earlier. The total particulate matter per cigarette comprises one fifth of the difference in weight of the Cambridge filter holders ($\pm$ 0.1 mg) before and after smoking, and it is the Cambridge filter disc plus the entrapped material that constitutes the sample for analysis.

3.1.5 Analytical methods

The Cambridge filter disc is transferred into a stoppered vessel containing a mixture of ethanol and propan-2-ol and shaken to extract the water which is then determined by gas-liquid chromatography. Five microlitres of the alcoholic extract is injected onto a Chromasorb 102 column at 160°C, and the eluent is monitored by a thermal conductivity detector. The four components, air, water (unknown), ethanol (internal standard), and propan-2-ol (solvent) are eluted in 3-4 minutes, and the quantity of water present in the trapped particulate matter is calculated from the ratio of the integrated peak areas of water and ethanol by reference to a calibration curve. This will normally lie in the range 0.1 - 8.0 mg/cigarette. To measure the total alkaloids (as nicotine) the alcohol extract is shaken with methanol until the glass fibre disc has completely disintegrated. An aliquot of the supernatant solution is transferred to a Markham still, containing 5 ml of dilute sulphuric acid. The acid-volatile interferences are discarded by distillation and strong sodium hydroxide is added to the still without interrupting the distillation rate. The distilled alkaloids are condensed into a flask containing dilute sulphuric acid, and the optical density (A) of this solution is measured against a dilute sulphuric acid blank, at 236, 259 and 282 nm in 10 mm silica cells. Alkaloids (as nicotine) are then calculated from a calibration graph prepared using solutions of nicotine hydrogen tartrate. Tar content is derived by deducting the water and alkaloids (as nicotine) values from the total particulate matter. The mean value of 30 such determinations per brand gives the published league table 'tar' value.

The distillation procedure for nicotine alkaloids is a very labour-intensive technique. With considerable physical effort and dexterity one operator can handle a bank of four Markham stills with a total throughput of 8 - 12 samples per hour. In addition, the spectrophotometer measurements at three wavelengths per sample are time-consuming when a manual instrument is used. When the techniques were set up in 1972, the initial steps in automation were taken with a view to easing the operator burden during the distillation stage and speeding up the measuring stage. The Markham still has some simple automatic functions in that it allows reagents to be introduced while the distillation rate is maintained, and it is self-emptying at the end of the distillation. A special steam generator was designed to simplify filling and emptying, and the distillation rate was controlled by a variable voltage transformer feeding the heating elements. The spectrophotometer was fitted with flow cells that allowed the samples to be pumped in and out of a silica cell sited permanently in the beam, and the wavelength selector was modified to locate quickly, positively, and reproducibly at 236, 259 and 282 nm.

The above procedures were required to handle some 30 000 cigarettes a year generating 7000 samples (Cambridge filter discs) for chemical analysis. The large quantity of data that this produced, all of which required recording, checking, collating, and reporting, needed either a large clerical team to support the analytical chemists or fairly sophisticated data processing facilities. The work was therefore started with a full computer-assisted data-processing support system, which provided the springboard for the subsequent total systems approach to automation.

3.1.6 Use of computer techniques

Software was designed by analytical chemists to utilize available computer facilities in four key areas: sample documentation, experimental design, data validation and reporting results. The operation of the program was dependent upon a coded list of all the cigarette brands being tested. Understanding the design philosophy behind the first brand code list is important in order to appreciate the key role that it played in co-ordinating, and constraining the changes that occurred as each stage of automation was implemented with the initial remote batch system as a foundation.

For the various cigarette brands, a dual coding system was devised which gave the analyst total flexibility of brand-code allocation, while preserving a rigid file accessing procedure. This dual coding system comprises an external code (the analyst's brand code), and an internal machine code (the computer file access code). The brand code is used by the analyst on all documents submitted to the computer for data processing and on all reports received from the computer. The access code (or, perhaps more properly, the access number, as it has no formal structure) is held by the computer in a table alongside the appropriate brand code, and it is used as a pointer to the record for the brand on the permanent data file. Alterations to the brand code have the effect of untying the brand code/access code link and retying the same access code to the new brand code. New brands have new access codes automatically allocated from a pool.

The brand code is structured so that the manufacturer, the type of cigarette and the size of cigarette are readily identifiable.

3.1.6.1 Sample documentation

The sampling procedure produces a regular flow of some 600 packets of cigarettes a month, all of which require careful documentation and labelling at the point of sampling, recording on receipt, and the transcription of all details onto the work sheets that accompany the samples during the various stages of chemical analysis. In addition, the sampling agency requires monthly instructions concerning the brands to be sampled at the various locations. The potential for errors is obviously large, and to minimize the clerical manpower needed to handle the documentation both by the sampling agency and the laboratory, the computer was used to combine all the requirements into one document in advance of sampling. This document consisted of sheets of self-adhesive labels printed on the line printer with all details required by the sampling agency and the laboratory. An identical pair of labels was printed for each sample required by the laboratory, and the complete set was sent to the sampling agency each month. These labels were used for sample identification and they also provided the sampling agency with a definitive list of samples to be drawn. After drawing the sample, the agency staff needed only to peel the appropriate pair of labels from the sheet, fix them to the cellophane wrapper of the cigarette packet, and, when all sampling was complete, despatch the entire set to the laboratory. This meant that there was no transcription. The documentation procedure at the LGC was equally simple; one of the paired labels was removed and fixed to a work sheet, leaving the other intact on the packet, so that sample and worksheet were exactly cross-referenced for all the subsequent stages in the laboratory.

3.1.6.2 Experimental design

One of the main reasons for determining the tar and nicotine yields of cigarettes is to give smokers a chance to choose a brand based on relative health risk. However, the analyst cannot provide absolute tar and nicotine values because these are dependent upon several variables, including the smoking habits of the individual. Accurate and precise relative values are required so that valid brand comparisons can be made. To achieve this, a statistically designed testing sequence was generated by the computer as a series of randomized smoking plans that covered the complete six months' set of samples (3500 - 4000 packets).

Initially, smoking plans were produced as a one-off print-out from a list of brands submitted to the computer staff by the analyst, but as the level of automation increased they were produced directly from the current brand file in which they were also stored. This provided the key to automatic routing of the data from the on-line analytical equipment to the correct position in the main result storage file.

3.6.3 Data validation

An essential requirement of any data storage and retrieval system is data validation to ensure that neither incorrect nor duplicate data are entered onto the main storage file which is used to generate the final reports. This preventative screen is required at four stages during the conversion of the cigarette sample into smoke chemistry values on the database.

1. *Transcription* of raw data from the analytical equipment onto worksheets.
2. *Calculation* of results from the raw analytical data.
3. *Preparation* of data from computer input.
4. *Addition* of data onto the computer file.

Data validation is very labour-intensive and expensive and is, at best, a coarse screen. A positive solution to the problem is on-line data capture, in which case both the data preparation stage and the need to recheck calculations are obviated. However, interaction by the analyst is still advised so that the data are validated prior to acceptance.

Calculation of the analytical results and subsequent checking of them by the computer provides a positive trap for transcription errors, and also provides an on-going check of the credibility of results. This often enables corrective measures to be taken without the need for complete re-smoking. All calculations are undertaken by the laboratory staff, and both the raw data and calculated values are submitted to the computer. These are then run through a program which calculates the results and compares these values with the results obtained by the laboratory staff. The worksheets corresponding to the error-flagged entries on the line printer output are re-submitted for punching, either with corrected calculations, or, if the error arose during data preparation, in an unamended form. This mutual checking procedure prevents incorrect data from reaching the main database.

Validated data are automatically written onto magnetic tape, whereas errors are ignored. At convenient times, which can be immediately, daily, weekly, or monthly, the data are loaded into the next available space in the appropriate record of the results file.

Figure 3.3 shows an extract from report on 'Tar, Nicotine and Carbon Monoxide Yields of Cigarettes' determined by the Laboratory of the Government Chemist for brands available during the period June to November 1994.

Tar Yield mg/cig	Brand Name	Nicotine Yield mg/cig	Carbon Monoxide mg/cig
10	Fine 120 Super Length Menthol	0.5	13
	Gauloises Caporal Filtre	0.6	13
	Gauloises Disque Bleu Caporal	0.6	13
	Gitanes Blondes Lights	0.8	11
	Haddows King Size Lights	1.0	9
	Kent King Size*	0.8	9
	Kingston Super Lights	0.8	11
	More Mild Menthol	0.9	10
	Red Band Lights Superkings	0.8	12

* indicates a low volume brand that is infrequently produced, the figures given are the LGC average yields from earlier surveys

Fig. 3.3   Extract from report on tar and nicotine yields of cigarettes.

A full report is required for the LGC's primary customer, the Department of Health; this is also made available to the cigarette manufacturers. At monthly intervals the database is manipulated, and results are exchanged between the original and re-test areas so that only a single valid result for each sample is available for the final reporting software. This final stage of data validation allows the operator to delete any incomplete record and to exchange matched entries in the original and retest areas. The analysts were thereby relieved of much of the clerical and administrative burden associated with calculation and collation, and the customer was provided with the league table directly in the form required for publication. An example of this is shown in Fig. 3.3. Complete analytical control was achieved only through full automation of the laboratory procedures with on-line data capture. This is described in the next section. Real-time analytical quality control was then a realistic goal.

3.1.7 Total systems automation

Automation of the laboratory procedures, the hardware linkage of these to the computer and the implementation of software to handle the incoming information were developed gradually, each being tested and proved in parallel with its routinely operating manual counterpart before being accepted as a working replacement for the manual procedure. Probably the most important stage in the development of the overall automation was a correct definition of the system before embarking on any hardware and software development. A schematic overview of the instrumentation configured for the smoking programme is shown in Fig. 3.4. Each section was developed as a stand-alone segment capable of being linked on-line as soon as the whole system was complete. The computer was the central controlling hub of the system.

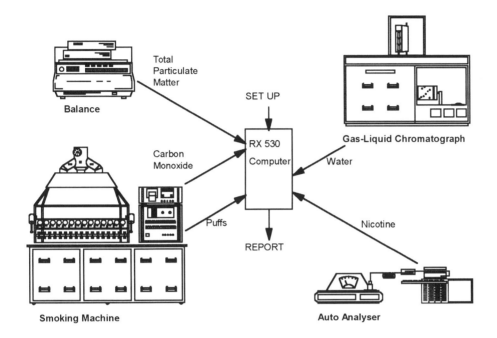

Fig. 3.4 Relationship between the fundamental parts of the total automation system. Data from all the analytical equipment is automatically acquired by the central co-ordinating computer.

The brand file and random smoking plans are central to the whole automation design philosophy. The brand file was the first segment of the system to undergo major revision when the data handling moved from the bureau to the in-house computer. The information on the brand file not only enables an analyst to monitor certain administrative requirements on behalf of the customer, but also allows the file itself to accomplish most of the control functions required by the data processing system. The software to drive this expanded brand file runs interactively from a terminal in the laboratory and gives the analyst the ability to update and maintain the file, and exert a measure of control over the whole system.

The brand file is used extensively by the validating, loading and reporting software, so its integrity must be beyond question. The software therefore provides security locks against damage to the file integrity, whether malicious or accidental, in the form of a file entry password, a confirmation request for any record entry changes made from a terminal, and the provision of an event log. Action by all programs that operate on the results file or the brand file are recorded in the event log, and an extended version of the program provides the analyst with the ability to interrogate this log.

The security and flexibility of the program can be illustrated by the following account of the manner in which it is used by the analyst. On loading, the program asks the user to type his name and a password. The name is used to mark all entries for the current session in the event log, and the password establishes the user's access rights to the file. There are three levels of access:

Level 0 Requires no specific password and allows the use of only help, list and exit commands. It is used only to trap and log illegal attempts to change the file.
Level 1 Requires the user password, known only to the supervisory analyst, and allows access to all normal commands that manipulate the file.
Level 2 Requires the system password, known only to the systems group, and allows access to certain privileged routines required for software development.

The program enters the command level when it recognizes a command. On recognizing a valid command it enters the option level and prompts the user for any further information it may need. On completion of the processing for the selected option, the program returns to the command level.

The password and event log provide very powerful protection against illegal attempts to gain access to the file, but do not prevent a user making incorrect entries. To provide a fair measure of safeguard against such inadvertent file corruption, particularly when the CHANGE options are being used, the program undertakes a sequence of checking routines that involve dialogue with the user. With the exception of PASSWORD, all command and option entries are sent back to the user's terminal for inspection. Before processing any CHANGE option, the entered alteration is sent back to the terminal as:

changing x to y OK?

Only after receiving an affirmative answer from the user will the alteration be written into the brand file. In addition, if the user enters the extra option VETO, the present entry in the particular field being manipulated is displayed on the terminal as:

option y for brand x is z, change it

Only when an affirmative answer is given will the dialogue to establish the change proceed; a negative answer immediately returns the program to the command level.

With a comprehensive interactive database for the brand established, it now became possible to integrate the sample documentation and experimental design programs so that the analyst could generate labels and smoking plans as required. Once the smoking plans were generated through the brand file they could also be stored in this file and thus became accessible to the results file, providing the co-ordination for handling information coming from automatic analytical equipment coupled on-line to the central computer system.

Of the manual methods, the most labour-intensive is the steam distillation/UV spectrophotometric method for nicotine alkaloids. Since this was the rate-determining stage of the entire analysis, it was clearly desirable to automate it first. The possibilities were to automate the manual steam distillation/UV spectrophotometer method for total alkaloids; to develop an alternative technique for total alkaloids based on a different chemical principle; or to install an automated gas-chromatographic method. This last option was the most attractive as it offered the prospect of being combined with the gas-chromatographic determination of water already in use. However, the use of gas chromatography necessitated changing the analyte from the total of nicotine alkaloids present to nicotine. Such a change required the consent of the medical experts, who, in fact, decided that the requirement for total nicotine alkaloids should remain in view of the uncertain physiological role of each member of the nicotine group of alkaloids, the varying proportion of these alkaloids in different types of tobacco, and the commercial and medical implications of changing the base against which epidemiological evidence is assessed. Therefore the gas-chromatographic approach was rejected.

Automation of the manual UV method using an AutoAnalyzer method that incorporates a flash distillation stage has been described [5] and has been used routinely. This technique is unsuitable for unattended automatic operation because the flash distillation stage requires constant supervision. Attention was therefore concentrated on alternative chemical methods of measuring total nicotine alkaloids. By far the most promising of these is an AutoAnalyzer method based on the Konig reaction [6]. The mechanism of this reaction has been discussed by Roy [7] and is illustrated in Fig. 3.5.

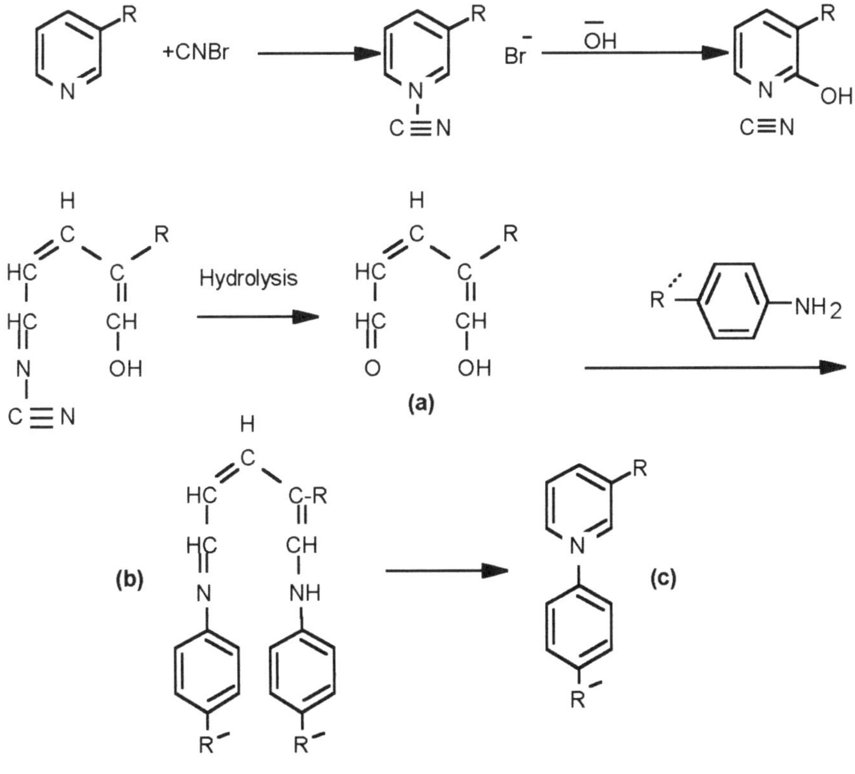

Fig. 3.5   Schematic representation of mechanism for colour formation in Konig's reaction (R = CONH$_2$ group, R = -SO$_3$H group).

A method using Technicon AAII equipment operating at 60 - 80 samples per hour has been demonstrated, and although not entirely suitable for this problem, it was a satisfactory basis for designing a procedure which could be interfaced to the RS-530 computer.

The method, shown schematically in Fig. 3.6, incorporates a dialysis stage to remove the alkaloids from a sulphuric acid extract of the condensed smoke trapped on the Cambridge filter disc.

The alkaloids pass through the dialysis membrane from the pH 4 buffered sulphuric acid extract donor stream into a pH 9 buffered aniline recipient stream. They are then reacted with cyanogen bromide to produce a yellow colour, the absorbance of which is measured at 460 nm after passage through a suitable delay coil to allow the full colour intensity to develop.

The automated procedure was designed to meet the following criteria:

1. The sample throughput should be 20 per hour.
2. The measurements should be made on the steady-state plateau, with a sample time sufficient to obtain a sensible reading.
3. The response should return completely to baseline.

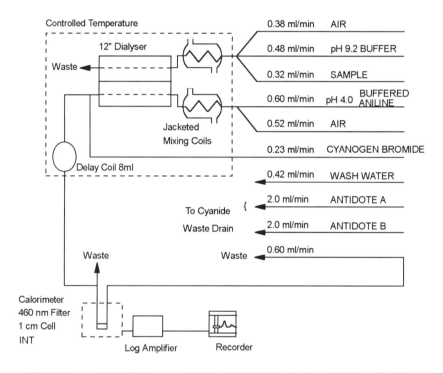

Fig. 3.6 Manifold diagram of nicotine alkaloids AutoAnalyser. Mixing coils, dialyser and delay coil are housed in a purpose built, light-tight controlled-temperature bath.

The instrumentation was constructed from a combination of purpose-built modules and equipment that was no longer in use. It consists of a Technicon (Bran & Luebbe) AAI sampler and 15-channel peristaltic pump, a constant-temperature path with purpose-built light-tight chambers to house mixing coils, colour development delay coils and 12-inch path dialyser using Technicon type C membranes, a much modified colorimeter fitted with a 460 nm interference filter, a purpose built logarithmic amplifier, and a potentiometric recorder. For safety reasons, the equipment was placed in a fume cupboard on a cambered tray that could be washed with water in the event of cyanogen bromide spillage.

Of the original design requirements, only the complete return to baseline between peaks was not met. For maximum peak width, a sampling rate of 20 per hour and a sample-to-wash ratio of 5:6 is used, and at these rates the baseline cannot be checked between samples. However, this disadvantage is overcome by the sampling protocol, which allows baseline drift and sensitivity changes to be regularly monitored and taken into account when the nicotine values for the samples are calculated. The sampling protocol also requires the insertion of a 0.5 mg/cigarette standard between each batch of four samples, and a water wash after each third standard. The baseline reading is taken before the initial standard, during each water wash, and after the final standard. The baseline value applicable to each sample and standard is calculated from the recorded baseline on either side of each set. An adjustment for sensitivity changes over each set of four samples is made by using the recorded value for the standards on either side of the set of samples that they bridge. The nicotine alkaloid value for each sample is calculated: the calibration is linear between 0 and 3.5 mg/cigarette and is checked by regular calibration.

Before being accepted as a replacement routine method, the AutoAnalyzer technique was extensively evaluated against the manual steam distillation method used at the LGC and against the various methods used by the UK tobacco industry. By replacing methanol with sulphuric acid at the second extraction stage, a solution of the nicotine alkaloids is obtained that can be examined by both the AutoAnalyzer and manual methods. Several thousand samples were examined, and despite the difference in chemistry of the two methods, the AutoAnalyzer was found to produce values not significantly different from the manual method. It was therefore adopted as the standard routine method, and it made a significant contribution to the laboratory's ability to increase its testing programme without an increase in manpower.

Attention was next turned to the weighing and water analysis procedures, which required a less radical change in methodology. Manual weighings were performed on a 0.1 mg mechanical balance, but to achieve on-line data capture an electronic balance suitably interfaced to the computer was required. A survey of the market revealed that a balance of this description with 0.1 mg precision was not available and automation could proceed only if 1 mg precision could be accepted. A statistical evaluation of the historical data was carried out to establish whether rounding the actual 0.1 mg weighings of total particulate matter from five cigarettes to 1 mg would significantly affect the final mean tar values. This showed that there would have been no significant difference in the reported tar values had the weighing been performed to only 1 mg precision. A Mettler PT200 balance and interface was therefore purchased. Like the nicotine alkaloids method, it is operated as a stand-alone system interfaced to a Silent 700 printer on which all weighings are printed. This provides an invaluable hard-copy record as an insurance against transcription errors. Experience with the top pan balance showed that manual weighing could be accomplished very quickly, and investment in designing and developing an automatic sample loader was not profitable.

For the water analysis, automation is clearly best achieved with an auto-injector for the mechanical handling of the samples coupled with on-line data capture, using the computer to analyse the peak data. Serious consideration was given to employing the very considerable in-house automation experience to construct a purpose-built auto-injector. However, in the interests of a speedy implementation of the automatic system, it was decided to purchase a commercially available auto-injector and to concentrate the laboratory's efforts on the area of on-line data capture. Interfacing the complete system assembly via a data communications network required the development of a special control device (commbox), which allowed the LGC hardware to run unattended but provided an audible warning in event of a fault condition.

The increased throughput capability of the AutoAnalyzer and electronic balance, coupled with the reduction in staff time, substantially increased productivity. Further benefits accrued because implementation of the automatic techniques also coincided with a need for the LGC to measure the carbon monoxide yields of cigarettes because of the interest in the carbon monoxide deliveries of cigarettes by epidemiologists working in the field of cardiovascular diseases. A development programme was initiated to find a method of measuring carbon monoxide yields simultaneously with the 'tar' and nicotine measurements, by monitoring the vapour phase components not retained on the Cambridge filter disc.

This method proved very successful in a single-channel development model [8], and plans that were made to scale up to a full 20-channel attachment for the routine smoking machine were discontinued only when a commercial carbon monoxide attachment for the Filtrona 300 smoking machine became available (ATCOM 20). This device automatically collects the vapour phase from each channel in a plastic bag, and at the end of the smoking run, the contents of each bag are sequentially exhausted through a non-dispersive infra-red carbon monoxide specific detector (NDIR). ATCOM was readily interfaced to the computer to allow channel-by-channel carbon monoxide concentration readings to be fed to it. The puff count, being displayed on mechanical counters, required a special interface.

Further stages were developed to:

1. Identify and accept the incoming asynchronous information from the six stages of analysis, all of which may be operating simultaneously with respect to different smoking plans.

2. Hold data and calculate final results, by reference to known calibration factors.

3. Report back to the laboratory at frequent intervals all important details of the analyses performed.

4. Continuously evaluate the quality of the analytical data being acquired, in relation both to the current run and to the expected values.

The fully automated system is shown in schematic form in Fig. 3.7. The controlling program BRANDER now contains the documentation (Labeller), experimental design (Planner), data manipulation (Alter), and reporting (RSVP) functions. Control of the analytical program is vested in the analyst through any of the terminals in the laboratory. The key to the fully automatic data handling is located in the ability to run Planner directly from the brand file, and thus have the complete six-month analysis schedule stored in that file. Each terminal operates any function of the automatic system and is not dedicated to a particular analytical instrument. The analyst can address a particular smoking run to any of the analytical stages, and the data produced will be directed to the appropriate record of the results file through the smoking run. Interactive setting up of a smoking run or data capture during analysis can take place simultaneously for several separate smoking runs.

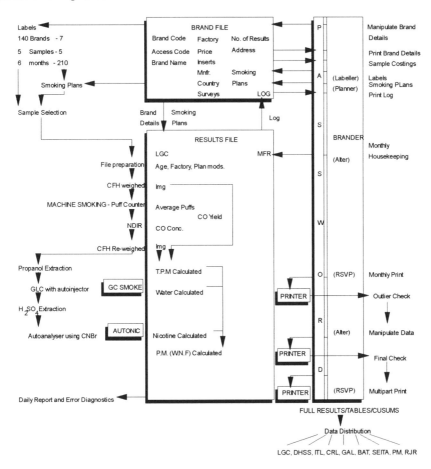

Fig. 3.7   Diagrammatic representation of the relationship between the automatic analytical procedures, the data files, and the large interactive control program BRANDER.

3.1.8   Summary

The most important task in any automation problem is to provide an accurate and detailed specification of the problem. The person most able to tackle this task is the analyst. Very often he or she will have little, if any, experience in the general area of automation and computing. This chapter has tried to illustrate how an analyst, in consultation with a multidisciplinary team of automation experts, has tackled a system for tar and nicotine analysis of cigarettes. Every facet of the analytical problem, from sampling through to reporting, has been automated to some degree. Open and honest co-operation has evolved a system which has not only solved the problem as it was originally presented, but has enabled the system to be expanded to cope with the changing needs. Whilst the technology for automation has advanced since this implementation was commenced, the lessons to be learnt are still valid. The overriding role of the analyst in the design of the necessary computing facilities cannot be underestimated.

3.2 CASE STUDY II - MERCURY ANALYSIS IN A NATURAL GAS PLANT

This second case study concerns the analysis of mercury in gaseous streams, particularly natural gas. Initially, analyses were to be carried out in the laboratory, but the final solution was to provide an on-line rapid analysis system. As in the first case study, the most demanding area of research and development was to find a suitable, reliable method of sampling the gaseous streams.

3.2.1 Introduction

The Minamata Bay disaster in 1953 (see Smith and Smith [9]) triggered an enormous amount of work, internationally, on techniques for determining mercury levels. Mercury occurs naturally in the environment in the form of mineral deposits and also anthropogenically from industrial and agricultural wastes.

In addition, mercury is often present in natural gas and petroleum products, both of which can be the basis of feedstock for industrial reactions. These are generally carried out with aluminium rotors or condensers, and low levels of mercury attack aluminium components, causing stress fractures. Mercury-induced corrosion on aluminium heat exchangers has resulted in at least four long-term industrial complex shutdowns in Algeria, the USA, Indonesia and Thailand. These shutdowns required costly replacements and from several weeks to several months of lost production; the plants have now been equipped with mercury removal units.

3.2.2 The problem

In 1987, an ethylene plant near Alvin, Texas, experienced an aluminium alloy piping failure which was attributed to mercury liquid metal embrittlement. English [10] has described the events that led to the failure and the action taken as a result.

The presence of mercury in the ethylene plant was first noticed when Algerian condensate was introduced as a feedstock during 1985. An in-line strainer directly upstream of the failure location began to plug. When an attempt was made to clear the pluggage, mercury was detected in the strainer. Analysis of the Algerian condensate showed mercury levels as high as 40 ppb by weight.

The problem was caused by cracks in a weld bead (see Fig. 3.8). This was discovered on a visual inspection (mercury wets the aluminum, giving it a tell-tale grey colour).

The piping failure was caused by the following:

(1) Periodic warming of the piping above the melting-point of mercury, resulted in liquid mercury coming into contact with the piping at the failure location.
(2) Erosion/corrosion of the weld bead at the fracture location damaged the aluminum oxide film on the piping, thus allowing the mercury to wet and initiate cracking of the aluminium.
(3) The cracks grew when the piping was warmed above the melting point of mercury.
(4) During nitrogen cooling, and the thermal cycling that occurred during nitrogen cooling, the applied stress on the piping was increased to about

10% above the stress experienced under normal, steady-state production conditions. These higher, cyclic, stresses caused the crack to grow into a through-wall fracture.

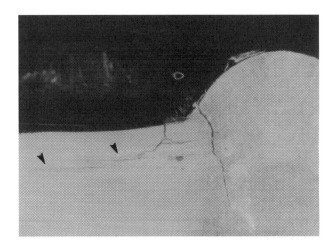

Fig. 3.8   Photograph showing crack in weld W-3.

3.2.3   Looking for a solution

According to Cameron *et al.* [11], mercury is redistributed into all of the hydrocarbon cuts after stream cracking. They found that the most highly contaminated fractions were butyl or propyl cuts, accounting for approximately 90 to 95% of the total mercury in the distillation column. About 75% of the mercury that entered a steam cracked unit was not found in the hydrocarbon products, but, rather, contaminated the internals of the unit. Of course, this mercury, when contacted with a clean, mercury-free feedstock will render the product mercury-rich. On the other hand, should the raw mercury-containing feedstock or one of its contaminated cuts be transported to another site, then the transport vessel will be polluted.

3.2.3.1   Mercury species

In order to identify the mercury species in the C3 and C4 cuts, experiments were carried out on propylene and on a raw steam-cracked C3 cut, containing methylacetylene (MA) and propadiene (PD). The steam-cracker C3 cut was found to be able to take up to 10 times more mercury than pure propylene. So MAPD has an effect on the mercury concentration in the C3 fraction.

It is unlikely that this is due to a simple difference in the solubility of elemental mercury in these cuts; it is more likely that there is a chemical interaction between the acetylenic species (ME and PD) with mercury. Elemental mercury and MAPD alone would not form Hg^{2+} organometallic compounds at room temperature, but an interaction between elemental mercury and the reactor walls (or impurities on the reactor walls) could well lead to the production of a more active form of mercury, which could react with MAPD to generate organometallic mercury.

Cameron *et al.* [11] concluded that a mild hydrogenolysis of raw feedstocks, or hydrocarbon cuts, will transform organometallic mercury into more easily trapped elemental mercury. Highly efficient, direct captation of elemental mercury can then be attained in the liquid phases with captation masses. A good captation mass for this purpose would be Procatalyse's CMG 273, (Procatalyse, Cedex, France) which has a highly macroporous structure; an anchored active phase which is insoluble in saturated, unsaturated and aromatic hydrocarbons; and very high thermal stability over a wide temperature range.

Mercury captation should be carried out as far upstream as possible in order to avoid contamination of process internals; regeneration of organomercury compounds after steam cracking; and the need to have a mercury removal unit on all of the process effluent streams.

3.2.3.2 Laboratory analysis

The example above shows the severity of the problem that arises from the presence of mercury in natural gas. Not only is it necessary to determine the levels of mercury present, but also to remove the majority of the mercury prior to any contact with aluminium reactors. The latter of course, further compounds the problem, because if 90% or more of the mercury has been removed, then to determine the remaining mercury is even more difficult. An effective analysis system will be able to measure mercury in its organic and inorganic forms and to do so very quickly. If a mercury removal bed is losing its efficiency then it is imperative to stop the process as soon as possible. In addition, these systems are expensive to operate and it is uneconomic to switch to a new unit if the original still has some life left in it.

There are many techniques used for the analysis of mercury in natural gas; the most commonly used approach is the Jerome Analyzer (Arizona Instrument, Phoenix, Arizona, USA). This collects mercury onto a gold adsorber over a period of time by amalgamation. The 431-X Mercury Vapour Analyzer is shown in Fig. 3.9.

Fig. 3.9 The Jerome 431-X Mercury Analyzer. *Photograph reproduced with permission of Arizona Instrument.*

Mercury vapour contained in a sample is drawn into the instrument by an internal pump. A gold film sensor adsorbs and integrates the mercury vapour, registering a proportional change in resistance, and displays the resulting signal on a digital meter. A microprocessor re-sets the analyser at the start of each sample and holds the resulting mercury value until the next sample is initiated. It detects from 1 to 100 nanograms of mercury; digital readouts are in nanograms with excellent reproducibility. The instrument provides a simplified laboratory procedure, which includes easy calibration and push-button operation for all instrument functions. It is designed for long-term performance without costly or complicated maintenance procedures. A major benefit of the Jerome analyser is that it eliminates expensive downtime for Hg analysis by cold vapour atomic absorption (AA) spectroscopy.

Comparisons between AAS instruments and the Jerome show statistically equivalent results. The Jerome gives accurate results for samples with a mercury content of less than 1 ppb and requires less work space, set up time, chemical reagents and sample size.

In addition, the AAS requires a calibration curve to be constructed to convert readings of absorbence to nanograms of mercury; whereas the Jerome provides direct readings in nanograms.

A problem with the Jerome is that it takes an extremely long time to collect samples if the mercury level is low, i.e., below 100 pg/m^3. In addition, it is prone to false readings and its ability to collect organomercury is suspect.

Another commonly used approach for the analysis of mercury is to pass the gas through a Dreschel bottle filled with potassium permanganate solution. Mercury is absorbed in this solution and then subsequently determined using the cold vapour technique. The efficiency of trapping is important and the flow must be set so that all of the mercury is collected. This is a time-consuming technique and it is prone to error, especially at the low levels where the background level in the reagent used can, in itself, become significant.

P.S. Analytical, in co-operation with Bayerisches Landesamt für Umweltschutz, has developed a dedicated system to analyse natural gas for low levels of mercury in both metallic and organic forms. Temmerman et al. [12] have described the collection of mercury and organic mercury species onto a gold-impregnated sand trap. P.S. Analytical has used similar traps in the design of the Sir Galahad instrument. Recent research has shown that a gold-impregnated silica support based on the spherisorb material provides significant advantages over the gold sand. Daniels and co-workers [13] have evaluated this material and its properties, and conclude that it offers significant advantages over other supports.

The Sir Galahad instrument takes advantage of the inherent sensitivity of the Merlin fluorescence detector. This detector was originally developed for the analysis of mercury in environmental samples and its principles have been described by Godden and Stockwell [14]. Stockwell et al. [15] have described specific applications of the device. It offers a more linear dynamic range, sensitivity, simplicity of design and the benefits of full automation.

Figure 3.10 shows the specific instrument design for trace mercury measurements. It provides a reliable on-line monitoring cycle, as well as the ability to analyse samples collected remotely from other sites. The system is described in more detail by Stockwell and Corns [16].

The unique capabilities of the system, including the method of absolute calibration, have been described by Corns et al. [17,18].

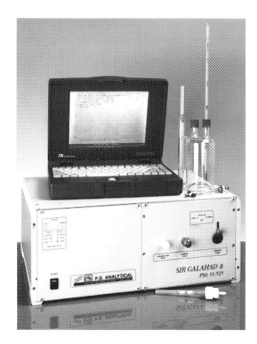

Fig. 3.10 The Sir Galahad System designed for the analysis of mercury vapour in gaseous streams.

The presence of air and other gases has the effect of quenching the fluorescence and reducing the detection levels attainable. To apply this detector to measurements in natural gas, some means of collecting the mercury and exchanging the gaseous media for argon is required. Dumarey *et al.* [19] developed a system utilizing gold-impregnated sand to collect mercury in air. Air is drawn over the trap at set flow rates for a set time, and the gold-impregnated sand collects the low levels of mercury present in the air. On completion of sampling, the trap is heated to revaporize the mercury into the detection system.

Figure 3.11 shows the collection process in diagrammatic form. This preconcentration device has been commercialized as the PSA Galahad preconcentrator or trapping instrument. The combination of this technology with the specific Merlin atomic fluorescence detector provides a fast, reliable, sensitive system and the necessary building blocks to design a tailor-made instrument to measure mercury in air and natural gas.

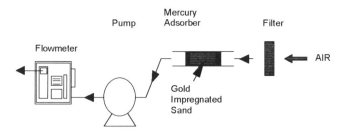

Fig. 3.11 Equipment for remote sampling of mercury in gas.

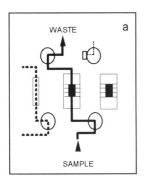 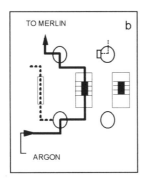 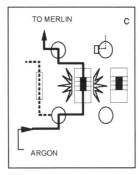

Fig. 3.12a,b,c   Configuration of Sir Galahad valves for collection, flushing, and analysis cycles in continuous mode.

To monitor continuously for mercury in natural gas, four cycles must take place. The first cycle involves pumping air or gas over the gold-sand trap to adsorb any mercury present (see Fig. 3.12a). The flush cycle (Fig. 3.12b) flushes out any gas that may be present, as this would have a damping effect on the mercury fluorescence. When this is complete the analysis stage begins. The heating coil is activated, which heats the gold sand to approximately 500°C. This desorbs the mercury from the trap so that it is carried in the stream of argon into the fluorescence detector (Fig. 3.12c). The fourth cycle is the cooling cycle, which simply pumps a flow of air or argon over the heating module to cool it rapidly, in preparation for the next loading stage. With the total times for the complete cycle ranging up to a maximum of 5 minutes, data can be returned within between 5 and 6 mins. This allows a continuous update on the mercury levels.

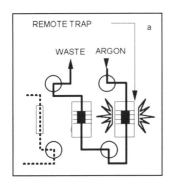 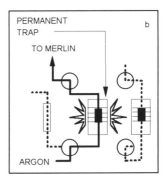

Fig. 3.13a,b   Configuration of Sir Galahad valves for transfer and analysis of remotely collected samples.

Remote sampling of gas not in the immediate vicinity of the instrument can be accomplished. This can be analysed by using specially designed removable gold-sand traps. Samples are first collected onto these using the method already described; the trap is then placed in the heating module, where it is flushed with argon to ensure that no traces of air or gas remain. The heating module then heats and vaporizes the mercury, which is carried by a stream of argon over the permanent gold-sand trap, where it is adsorbed (Fig. 3.13a). The permanent gold-sand trap then goes through its heating cycle and the mercury is revaporized into the fluorescence detector, where a response is produced (Fig. 3.13b). This system ensures that consistent results are returned, as the permanent gold trap is used to calibrate the instrument.

A typical response is shown in Fig. 3.14. All operations of the instrument can be controlled by the TouchStone© software [20], which will run on an IBM compatible computer. The calibration procedure is simple, effective, and runs automatically, prompted by the software. This involves the introduction of mercury-saturated vapour at a set temperature directly onto the permanent analytical gold-sand trap. Elemental mercury used as the primary standard is contained at atmospheric pressure in a specially designed glass vessel, held in a thermostatic bath. The temperature is monitored using a thermometer. A set volume of mercury-saturated vapour is drawn into a syringe from the vessel and then injected directly into the argon stream onto the permanent gold trap through a septum fitting. The amalgamated mercury is then vaporized and transferred into the detector for measurement.

The saturation concentration is calculated by the software. The signal from the Merlin shows a transient peak which is measured by peak height. In addition, the time of the peak maximum is recorded, providing further identification of the mercury present. A linear calibration is obtained from which analytical results are calculated. The software is also available to couple this system directly to a process control computer.

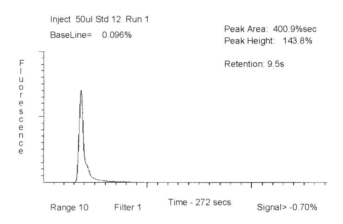

Fig. 3.14   Typical transient signal from the Sir Galahad instrument.

The system provides a very sensitive means of detection: levels of 10 picograms absolute are measurable with the continuous (permanent) trapping system. A further advantage is that the software calculates the analytical results directly in concentration in the unit volume of sample introduced. However, it should be stressed that the level of mercury measured is an absolute quantity and while the detection limit is of the order of 10 picograms, this quantity can be contained in any volume of gas. In addition, the fact that the mercury both absorbs and fluoresces to provide a measurement which can be measured with a specific retention time provides more positive evidence of the presence of mercury.

The Sir Galahad has been found to provide rapid, reliable analysis, often in minutes, while other techniques rely on a long sample-collection timescale. Recent developments in the choice of adsorber have provided additional benefits, mainly related to reduced back pressure and less thermal lag. These allow the analyses to be completed in shorter timescales.

The introduction of the Sir Galahad has allowed a more reliable total systems approach to the problem of mercury measurement.

The Sir Galahad has been used in many locations around the world. For example, Cameron et al. [11] have used the instrument to validate plant operations in Thailand. In this application the Sir Galahad was compared initially with other systems. Acidic vapour was found to cause false positive results in the other systems, whereas the Sir Galahad provided consistently reliable data. Tests have recently been carried out to compare the Sir Galahad with conventional techniques - Table 3.1 shows results produced using the Sir Galahad and the ISO 6978 potassium permanganate absorption method.

Table 3.1 Comparative results for the Sir Galahad and ISO Standard 6978 techniques

Sample	Potassium Permanganate Absorption Method ISO 6978			Sir Galahad Method ISO 6978 (Amended)		
	Volume Gas (litre)	Sampling Time (min)	Mercury ($\mu g\ m^{-3}$)	Volume Gas (litres)	Sampling Time (min)	Mercury ($\mu g\ m^{-3}$)
1	24.4	30	108.7	0.513	1	118.6
2	47.6	60	71.9	0.997	2	78.1
3	71.9	90	*	4.747	10	37.5

* Oversaturated

3.2.4 On-line analysis

Figure 3.15 shows a schematic overview of a natural gas plant and indicates sampling points that may require analysis. These are indicated prior to, between and after the mercury removal beds.

Originally, the plan was to introduce a sophisticated gas sampling system which would analyse samples at each of these sampling points. In order to achieve this, a pilot sampling system was designed such that the volume of gas sampled for the various stages of the process would effectively yield enough mercury to provide a repeatable signal. All these samples were to be taken at a pressure of 83 atmospheres. The volume of sample used for each range was therefore set by using set loop sizes. As the samples are collected onto the gold trap in the instrument, the loop sequence is saturated so that the gas samples are moved over the trap by natural expansion and flushed through the trap with a flow of argon or natural gas. Once the gas sample has been passed over the trap, the analytical cycle proceeds in the normal manner. All of the samples are collected by using a standard speed loop arrangement and a series of valves.

Although the sampling scheme above provides a viable theoretical solution to the problem, a natural gas plant is an enormous installation covering a large area of ground. The pipelengths involved in such a scheme would be quite daunting. Initial trials using gas bombs to collect representative samples and bring them to the laboratory area showed that this was a way of identifying the most critical sites for monitoring. Only one site required continuous monitoring to provide a constant update on the performance of the removal system and to provide protection of the plant. This was the site directly prior to the input into the process stream. Mercury levels at this point provide a protection against the introduction of mercury into the process line, and serve to monitor the performance of the captation mass. An advantage of the Sir Galahad instrument is that it does not provide false positive or false negative results. Both results would be expensive or even disastrous to the economic operation of the process plant.

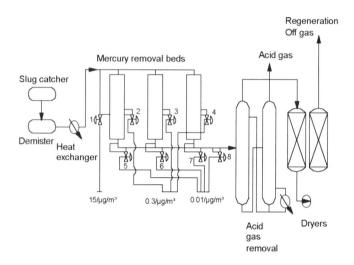

Fig. 3.15   Schematic overview of a natural gas plant.

An appropriate solution to the automation of this project was therefore to link an analysis instrument directly on-line to the process at this point. This provides rapid analysis and updating of the results on a continuing basis. Since the instrument was designed as a laboratory instrument, it is essential that it is correctly protected and located. This is achieved by placing it in a nitrogen purged instrument workstation. In this way it can be linked close to the process line.

The detailed information provided by the measurement of mercury is vital to the economic operation of the plant, so it is necessary to link the measurements directly into the process control computer. This is achieved in a simple manner by adding a digital-to-analogue converter to the control computer. Once the analytical result is calculated for any run, the result can be transferred directly by the D/A converter to the process control computer in a 0-20 mA signal.

In addition to analysing samples directly on-line, the instrument can be used for routine tests of samples from other sampling points. These samples are brought to the instrument after collection in suitable gold traps. Care must be exercised in this type of measurement, as with all others for mercury - the levels of measurement are very low and background levels can distort the result.

3.2.5   Conclusion

The availability of a new highly sensitive system for the measurement of mercury in natural gas has allowed a completely new total systems approach to be adopted. This provides an on-line system to measure the most critical gas stream on a continuous basis, sending detailed results directly to the process computer. It avoids the built-in delays involved in using laboratory measurements and provides better integrity of results.

3.3 CASE STUDY III - WATER LABORATORY AUTOMATION

3.3.1 Introduction

This case study looks at the way that a UK water laboratory service has automated testing in order to deal with increasing legislative and regulatory pressures. Severn Trent Laboratories (STL), the subject of the case study, currently operates from four sites and is the largest laboratory service in the UK water industry, handling 250 000 samples and 1.75 M determinands a year. This case study explains the challenges faced by the service and the work of the two main laboratories, that analyse potable and waste waters. This section is written in co-operation with staff from Severn Trent Laboratories, Birmingham Laboratory, 156-170 Newhall Street, Birmingham B3 1SE, UK.

3.3.2 Inorganic water analysis

The laboratory is highly automated, specializing in the analysis of large numbers of water samples for a wide range of parameters. It receives an average 650 samples each day of potable water, sewage, and industrial wastes, on which it performs a total of 5000 determinands daily.

Wherever possible, analytical systems are automated and directly interfaced to a laboratory data system. This mainly receives information from an Optical Mark Registration (OMR) system, or from down-loaded computer files, and then produces a work schedule which is printed for the analyst and transferred electronically to the instrument.

On completion of an analysis, the results are checked and validated by the analyst and compared with preset limits based on permitted concentration values or consented values, before being sent electronically to the database. Any result which breaches a particular limit can be repeated for confirmation. About 80% of the work is fully automated in this way, with most of the major items of equipment being interfaced to the Laboratory Information Management System (LIMS).

3.3.2.1 Automation

Examples of the automated systems in operation are:

(a) Three linear robots which analyse between 300 and 400 Biochemical Oxygen Demand (BOD) tests daily. BOD measures the biochemical oxidation of organic matter by heterotrophic bacteria over a five-day period. This technique is used to assess the biodegradable nature of samples.

(b) Three hundred Chemical Oxidation Demand determinations (CODs) are performed daily on settled samples using an anthropomorphic robot for the titrimetric finish.

(c) Over 1500 metals are determined daily by the techniques of atomic spectroscopy, including flame and ETA-AAS, AFS and ICP-OES and ICP-MS.

(d) Nutrients, such as ammonia, total oxidized nitrogen, nitrite, chloride, phosphate, sulphate, and silicates, are determined by using three discrete analysers, each of which is capable of a throughput of 600 determinations per hour.

(e) Fully automated equipment is also available for the determination of physical parameters, such as pH, electrical conductivity, colour and turbidity.

Analysts at STL are constantly being challenged, whether it be to improve a method so that the reporting limit can be lowered or to adapt a method so that it is suitable for automation.

3.3.2.2   First challenge: the legislation

Following a recent World Health Organization (WHO) recommendation to lower the permissible level of lead in drinking water from 50 µg l^{-1} to 10 µg l^{-1}, methodologies were required to detect a tenth of the new limit, i.e. 1 µg l^{-1}. Work was carried out at STL to introduce ICP-MS; the limit of detection (LOD) is much lower by this technique than by furnace. Good results for Cd, Cr, Cu, Pb, Ni, Zn and Ag have been achieved by ICP-MS using the following three internal standards: Sc (45), In (115) and Ir (193). This technique is now used in the routine determinations of low-level metals in potable water samples.

3.3.2.3   Second challenge: reduction of analysis time

Currently, water quality is assessed by the measurement of several parameters such as BOD, COD and ammonia. There is a delay time associated with such parameters: for instance, determination of BODs requires a five-day incubation. New techniques are being developed, for example enhanced chemiluminescence. This is a simple, sensitive and rapid assay which can be used as a diagnostic tool for monitoring water quality. The technique involves the oxidation of luminol catalysed by horseradish peroxidase. The reaction emits light at a constant rate. Many compounds in waste waters act as radical scavengers, and these, with enzyme inhibitors such as cyanide, phenols and heavy metals, cause a reduction in chemiluminescence. The reduction and shape of the recovery curve can provide information on the concentration and type of pollutant present. Results show high correlation coefficients with BOD and COD measurements. Applications could include supply intake protection and assessment of river, sewage effluent and bathing-water quality.

3.3.2.4   Third challenge: new automation

The established method for phosphorus analysis using a heated block involves a lengthy digestion with sulphuric acid. Furthermore, it is labour and space intensive and requires the digested sample to undergo a second stage for the final analytical determination.

Recently, work has been undertaken to develop an on-line microwave analyser. Acidified sample is injected into a carrier stream which passes into a microwave oven. During the digestion period of 1 min, phosphorus-containing compounds are converted into the orthophosphate form. After cooling, reagents are added to form phosphomolybdenum blue, which is measured spectrophotometrically at 690 nm.

The system is linear up to 40 mg l^{-1} of phosphorus and has an RSD of less than 5%. This automatic system surpasses manual batch methods in both precision and bias.

3.3.3 Sludge, soil and organic water analysis

A range of routine analyses and specialist services are also provided for samples of sludge, soil and organics in water. More than 60 types of pesticides/herbicides are quantified daily.

3.3.3.1 Sludge and soil analysis

The laboratory carries out routine analyses of sludge and soil samples for nutrients and metals before application to agricultural land. Severn Trent is the only land-locked UK water company and so, unlike the others, has no direct access for disposal of sludge to sea (the Urban Wastewater Treatment Directive has now curtailed this disposal route). Around 10% of the sludge is incinerated, with most of the remainder being applied to agricultural land. The sludge is analysed for nutrients and 10 selected metals. Also, the land on which it is to be applied is sampled and tested before the sludge is spread, in order to ensure correct application rates.

3.3.3.2 Organics water analysis

A large group is devoted to the determination of organic compounds in water and other materials. The team of over 40 scientists uses the latest technology - for example gas chromatography, gas chromatography coupled with mass spectrometry, high performance liquid chromatography and high performance chromatography coupled to mass spectrometry - to analyse for a wide range of insecticides, herbicides, fungicides, phenols and chlorophenols, haloforms, chlorinated solvents and polycyclic aromatic hydrocarbons.

With the introduction of the UK Water Supply (Water Quality) Regulations, 1989, the water industry has been required to develop a monitoring strategy for pesticides, based upon usage and the risk of contamination in a particular catchment area. Since the limit is 0.1 $\mu g\ l^{-1}$ for individual pesticides and 0.5 $\mu g\ l^{-1}$ for total pesticides, the challenge to the analyst is ever-decreasing detection limits for an ever-increasing range of compounds. Analysis of such concentrations presents considerable technical problems, especially within a routine environment. Breaches of the limits can impose further sampling by the water company and the results have to be made available through the public register. Thus, it is essential that the analyses should be of the highest quality to ensure appropriate decision-making based on the results.

3.3.3.3 Sample preparation

Because of the low detection limit requirements, a concentration factor from the usual 1 litre of sample of between 1000- and 10 000-fold is required. This is usually achieved by solvent extraction, followed by evaporation of the extract to a small volume. However, the introduction of the Control Of Substances Hazardous to Health Regulation, 1989 (COSHH) has caused concern about the relatively large volumes of solvent used for extraction - especially chlorinated solvents such as dichloromethane. In order to reduce the volume of solvent used, solid phase extraction, using commercially available C_{18} bonded-phase cartridges, is being introduced.

3.3.3.4 Analysis

The main analytical technique for pesticides is gas chromatography using a variety of detectors such as electron capture for halogenated compounds, thermionic for compounds containing either phosphorus or nitrogen, and flame photometric for compounds containing sulphur or phosphorus.

For separation, high resolution capillary columns are used, and in most circumstances each extract is examined by a dual column technique. This is to add a degree of confirmation into the method.

High-performance liquid chromatography is also used, for the analysis of thermally labile compounds. The laboratory has a range of detectors, although most work is carried out using UV or diode-array detection. Diode-array detectors have been used increasingly in preference to UV detectors. Because of possible interference from other sample compounds, retention time is insufficient as the only criterion for positive identification. Results acquired using diode-array detectors in conjunction with an automated library search programme have been successfully used to screen samples for various pesticides including carbendazim, carbetamide and difenzoquat.

Although diode-array detectors offer more confirmation than single wavelength UV detectors, it has been found necessary to use mass spectrometry coupled with HPLC for certain determinands. At low levels for certain compounds, spectra obtained from diode-array detectors are not fine enough for confirmation purposes. Groups of compounds may have similar spectra and so co-eluting peaks may not be correctly identified. Uron herbicides are one such group. Compounds such as isoproturon and diuron have been found in waters above the limit of 0.1 $\mu g\ l^{-1}$ using UV detection. The only method that can identify these compounds with certainty is LC-MS. From all the presumptive uron herbicides, less than 50% were confirmed by LC-MS and there is still much work to do in confirming other compounds by LC-MS.

With a combined throughput of around 250 000 samples and over 1.75 M determinations a year, quality assurance is of paramount importance. This is ensured by a continuous process of introspection and audit, both internal and external.

The key areas are as follows:

(a) *Accreditation*

STL are accredited by the Department of Trade and Industry via the National Measurement Accreditation Scheme (NAMAS). This covers all aspects of laboratory operations, such as organization of the laboratory and methodology, equipment and staff training, quality-control systems and storage of data. The company acquired the first accreditation in the UK for the analysis of organic compounds by GC-MS. It also operates appropriate procedures to conform with the Department of Health's Good Laboratory Practice (GLP) recommendations.

(b) *Performance*

The performance of each method has to be established so that the precision and limits of detection can be derived. The Water Supply (Water Quality) Regulations 1989 specify acceptable limits of deviation and detection for individual methods. Methods for potable water can only be used if they conform with these specifications. These are achieved by running the following solutions in a random order each day over a 10-day period. Once calculated, the maximum tolerable errors of individual analytical results can be defined as:

i. Total error of individual results should not exceed 10% of the maximum allowable concentration (MAC) or 20% of the result, whichever is the greater.

ii. Total standard deviation of individual result should not exceed MAC/40 or 5% of the result.

iii. Systematic error (or bias) should not exceed MAC/2 or 10% of the result.

(c) *Quality control*

The laboratories have a quality manager and quality section. The quality section prepares all of the analytical quality control (AQC) standards for the laboratories, including spiked and duplicate samples. The quality section is a separate laboratory with a separate supply of deionized water, glassware, balances and chemicals. The latter, wherever possible, are purchased from a different source to those for the analytical sections. These actions ensure a more independent approach to quality control.

STL's quality-control programme includes the recovery of known additions of analyte, analysis of externally supplied standards, calibration, analysis of duplicates and control charting. Each analyte is monitored by analysing at least one AQC standard for every 20 samples. AQC results are plotted on control charts and action is taken if a point lies outside ± 3 standard deviations (SD) or if two consecutive points lie outside ± 2 SDs.

In addition, results which breach these limits are recorded in logs and repeat analyses are carried out. Results are accepted only when they conform to all these stages of AQC.

The laboratory is also involved in quality assessment, such as participation in an external quality-control scheme (inter-laboratory comparison). Here, a variety of analytes in matrices such as hard water, soft water, and sludge are circulated to laboratories and the returned results are processed in a codified format so that each participating laboratory can only identify its own values.

Quality is further maintained by external inspection. The laboratories are inspected by a number of agencies on a yearly basis, for example, by the Drinking Water Inspectorate. Under the UK Water Act 1989, the Secretary of State for the Environment and the Secretary of State of Wales need to satisfy themselves that water companies are supplying wholesome water and that they are monitoring, recording and reporting on the quality of drinking water in accordance with the legal requirements. Under Section 60 of the Act, technical assessors are appointed to inspect every aspect of water supply, which includes analysis of samples. The audit is detailed and comprehensive.

The laboratories are inspected annually by the National Measurement Accreditation Scheme (NAMAS) and for conformity to GLP. Part of the inspection is a vertical audit, which involves tracing a sample from the moment it is delivered to the laboratory to when the result is reported. All the records of its passage through the system are traced. For NAMAS, all records need to be kept for six years; for GLP the time period is as agreed with the customer.

Also, the laboratories can be inspected by their customers. A range of customers, from large companies to consultants, avail themselves of this opportunity in order to satisfy themselves of the validity of the information that they receive.

STL has relocated some of its operations. It has expanded its Coventry Finham site to set up a clean water laboratory and its Coventry Tile Hill site for liquid and solid wastes. These sites now form the Midlands operation of STL.

REFERENCES

[1] *Smoking and Health Now*, Report from the Royal College of Physicians, (Pitman Medical and Scientific Publishing Co.), 1971

[2] Research Paper No. 1: *Standard Methods for the Analysis of Tobacco Smoke*, Tobacco Research Council, London 1972 and 1974. Obtainable from Tobacco Advisory Council, London.

[3] ISO 3402-1975, International Organization for Standardization.

[4] ISO 3308-1977, International Organization for Standardization.

[5] Ayers, C.W., *Technicon Symposium Automation in Analytical Chemistry*, Mediad Inc, NY, 1967, 1966, Volume II, 107.

[6] Konig, W., *Journal für Praktische Chemie*, 1904, 69, 105.

[7] Roy, R.B., In *Topics in Automatic Chemical Analysis*, Foreman, J.K. and Stockwell, P.B. (Eds.)., Horwood, Chichester, UK, 1979, p. 256.

[8] Browner, R.F., Copeland, G.K.E., Stockwell, P.B. and Bergman, I., *Beitrage zur Tabakforschung*, 1977, 1, 9.

[9] Smith, W.E. and Smith, A.M. (Eds) *Minamata*, Holt, Rinehardt and Winston, New York, 1975.

[10] English, J.J., A.I.C.L.E. Spring Natural Meeting, Houston, USA, 1989.

[11] Cameron, C.J., Sarrazin, P., Barthel, Y., Courty, P., Shigemura, Y. and Hasegawa, T., 3rd International Petrochemical Conference, Singapore, 19 September 1991.

[12] Temmerman, E., Vandecasteele, C., Vermeir, G., Leyman, R. and Dams, R., *Analytical Chimica Acta*, 1990, 236, 371.

[13] Daniels S. and Vermeir, G., Personal communication, 1993.

[14] Godden, R.G. and Stockwell, P.B., *Journal of Analytical Atomic Spectrometry*, 1989, 4, 301.

[15] Stockwell, P.B., Henson, A., Thompson, K.C., Temmerman, E. and Vandecasteele, C., *International Labmate*, 1989, 14, 45.

[16] Stockwell, P.B. and Corns, W.T., *Spectroscopy World*, 1992, 4, 14.

[17] Corns, W.T., Ebdon, L.C., Hill, S.J. and Stockwell, P.B., *Journal of Automatic Chemistry*, 1991, 13, 267.

[18] Corns, W.T. and Stockwell, P.B., *Hydrocarbon Asia*, 1993, Oct, 36.

[19] Dumarey, R., Temmerman, E., Dams, R. and Hoste, J., *Analytica Chimica Acta*, 1985, 170, 337.

[20] TouchStone Software©, SpinOff Software, Copyright of SpinOff Technical Systems, Benfleet, Essex, UK.

4

Automatic sample-preparation and separation techniques

4.1 INTRODUCTION

Sample preparation is of paramount importance in an analytical laboratory, but automatic sample preparation has had relatively little attention. Developments have often simply been mechanized versions of standard analytical procedures. When an analysis is automated it is essential to look at the overall analytical problem - it is important initially to determine whether or not sample preparation is required (each analytical step which is automated complicates the instrumental requirements, making it increasingly difficult to obtain reliable repeatable analytical results).

In a working analytical laboratory, a considerable percentage of the analyst's time will be spent on sample preparation. A range of analytical techniques will be used, such as solvent extraction, filtration, digestion and flash distillation; chromatographic techniques have received a great deal of attention and the instrument companies have introduced automated systems. Two widely differing approaches have also been introduced - the use of infra-red reflectance techniques and the application of robotics (these techniques will be discussed in detail later). When a method is transferred from a manual regime to an automatic system, it is important to give a correct specification of the analytical needs. A simple approach is to mechanize a manual procedure directly. This is not always the best approach, however, because new technology and the use of automated systems can allow alternative, more easily automated procedures, to be adopted. Whenever an automated method is introduced, whether or not it uses a similar approach to that used manually, it is vital that the full procedure is correctly and properly validated (as shown in Fig. 4.1). Automating the procedure can subsequently provide a great deal more credibility to the measurements being made.

This chapter outlines many techniques which have been used directly by the author. Whilst the content does not provide a complete coverage of the available literature, it does serve to highlight many of the attempts made to automate this difficult area. It is hoped that it will serve to prompt clear thought on the part of the system designer and also point out some of the difficulties and dangers incurred.

4.2 LIQUID - LIQUID EXTRACTION (SOLVENT EXTRACTION)

Applications of solvent extraction techniques are numerous in the scientific literature; but there are few, if any, good commercially available automatic systems. To ensure good, reliable results in an automatic system, it is important that the analyst has information on the relative solubility of the compounds of interest in the respective phases. In a manual regime, properties that are less than ideal can be catered for by multiple extractions and backwashing. However, in an automatic system these parameters must be built into the instrument specification, i.e. the procedures must all cope with the greatest variation of sample type or must be flexible enough to change from one sample to another. For example, a microwave extraction procedure which has a tendency to foam for some samples must be set up to avoid foaming and this may create an over-long time frame for a set of analyses.

Solvent extraction can be automated in continuous-flow analysis. For both conventional AutoAnalyzer and flow-injection techniques, analytical methods have been devised incorporating a solvent extraction step. In these methods, a peristaltic pump delivers the liquid streams, and these are mixed in a mixing coil, often filled with glass ballotini; the phases are subsequently separated in a simple separator which allows the aqueous and organic phases to stratify. One or both of these phases can then be resampled into the analyser manifold for further reaction and/or measurement. The sample-to-extractant ratio can be varied within the limits normally applying to such operations, but the maximum concentration factor consistent with good operation is normally about 3:1.

It is important to design the phase separator correctly and to adjust the rate of removal of the respective phases from it. In the case of aqueous determinations, globules of an organic phase entering the flow-through cell will give a false reading. A far greater limitation is presented by the nature of the pump tubes; but recent developments in the plastics industry and the use of displacement techniques can overcome some of the problems [1]. The Technicon Evaporation-to-Dryness Module can also extend the concentration factors to more acceptable levels [2]. In this, samples extracted into a volatile solvent are placed on a moving inert belt over which air or vacuum is applied. The solvent evaporates and the sample is then re-dissolved in another solvent as the belt moves into a new section of the manifold. This technique is particularly suitable where there is a need to change the solvent matrix to ensure compatibility with the measurement stage, as in liquid chromatography. The application of solvent extraction in flow-injection applications has been described by Karlberg and Thelander [3].

There are various ways of controlling the phase separation stage in discrete analytical systems; both static and dynamic systems have been described. A dynamic approach using centrifugal force was developed by Vallis [4]. Vallis's device comprises a cup-shaped vessel mounted on a rotor complete with a porous lip attached to the lip of the cup. The device is placed inside a collecting vessel, and if the porous lip is made from a hydrophilic material, such as sintered glass, water will pass into the collecting cup at low rotation speeds, leaving the organic phase retained in the cup. An increased rotation speed is then used to reject the organic phase. A hydrophobic interface such as sintered PTFE will allow the organic phase to be rejected. A major problem encountered by such systems is that the interface is unstable and requires frequent replacement or regeneration, thus negating many of the advantages of automatic operation.

Centrifugation to aid an operation has found considerable use in the General Medical Science/Atomic Energy Commission approach to automation pioneered by Anderson [5], and was also used by Bartels *et al.* [6], in the design of a solid-liquid extraction system.

A prototype separator, designed and built by Williams *et al.* [7], using a completely new approach, was the subject of a patent application in 1980. In principle, separation is effected by absorption of both phases into a porous nickel-chrome alloy disc mounted on a motor-driven shaft. Controlled angular acceleration and centrifugal force on the droplets within the pores enables one phase to be separated from the other. The speed of rotation of the porous disc is coupled microelectronically to the vertical component of its motion so that separated droplets leaving the disc tangentially are trapped by hitting the walls of concentrically arranged glass vessels. Valves are provided at the base of each system so that separated droplets may be removed for further processing. By applying a potential between the rotating disc and a rigid electrode situated about 5 mm from the edge of the rotor, a current can be sensed as soon as the speed of rotation has increased sufficiently to cause spin-off of liquid droplets. This signal may be used to instruct the motor to continue to run at a constant speed.

The apparatus is shown in Fig. 4.1. The body of the extraction vessel is made from Pyrex. Separation is effected by absorption of a batch containing both phases into a porous 2 cm diameter nickel-chrome alloy disc (A), the upper surface of which is domed. The disc is mounted on the end of a stainless-steel shaft (B) turned by a geared high-torque electric motor. The disc-shaft-motor assembly can be transported along its axis of rotation to any of three stations. The assembly is shown at its bottom station, with the porous disc within the inner vessel (C), around which is a collar (D) forming the first annular pocket (E). The collar itself forms the inner wall of the second annular pocket (F), the outer wall of which extends upwards to support a Perspex lid (G) on which the rotor (H) is situated. The inner vessel and both annular pockets are fitted with drain valves. A stiff piece of platinum wire is passed through the lid into the glassware as far as the level of the first annular pocket.

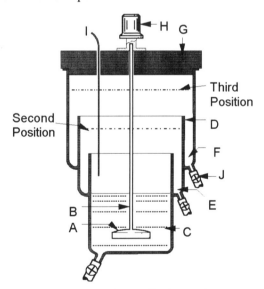

Fig. 4.1   Schematic layout of a centrifugal separation system.

In operation, the mixed liquid to be separated is pumped into the vessel, covering the disc at its bottom station. The disc is set to spin at high speed, thoroughly mixing the liquid. The spinning of the disc is then stopped and the disc is raised electromechanically to a position just above the top of the upstanding collar. At the same time, the motor starts to spin the disc, the speed of which is smoothly increased until droplets of the first phase come off and a significant current flow is observed between the rotating disc and the platinum wire. The rotor continues to spin at constant speed for 15 seconds, sufficient time for the first phase to be thrown off the disc. The disc is then raised to its top station and accelerated, throwing off the aqueous phase. The rotating disc remains in this position for a further 15 seconds, after which it returns to the lower position. The process is then repeated. The linear electromechanical actuator and motor used for spinning the disc were both obtained from Portescap (UK) Ltd. Either phase may be selected for 100% purity by adjusting the sensitivity of the droplet detector. In general, the second phase is 70 - 75% pure. The apparatus has been used with several solvent combinations, including chloroform/water.

The nickel-chrome alloy is available from the Dunlop Aviation Group and has the trade name of 'Retimet'. Retimet is suitable, although its affinity for the organic phase is greater than for the aqueous phase. In order to produce approximately equal affinity for each phase the discs were gold plated for 10 min using a current of 300 mA. It is preferable that the alloy is shaped so that the pore structure is maintained on the surface after machining. Spark erosion is used to shape the material.

In static solvent-extraction devices, a variety of phase-boundary sensors controlling simple on/off valves to govern the flow of the liquid through them are used to effect the solvent separation. An early system, using a pair of conductivity-sensing electrodes and a pinch-clip valve, was described by Trowell [8]. It is not advisable to use such a device when volatile flammable solvents such as ether are required. In such cases there is a potential explosion hazard should a spark occur between the electrodes. Phase-boundary detectors with sensors based on conductivity, capacitance, and refractive index can be used. These detectors have been reviewed by Stockwell [9]. The detector should ideally be external to the organic/aqueous phases. A compromise choice must inevitably be made between retaining an unwanted small portion of one of the phases and the loss of a small fraction of the phase of interest. In practice, the speed of stirring is of considerable importance; too rapid a speed could generate an emulsion from which either fails to settle, or settles after only a prolonged time. Knowledge of the chemistry involved in the analytical procedure, and the aims of the analysis, are required to optimize the instrument design. No single type of phase-boundary sensor is universally applicable.

4.3 FLASH VAPORIZATION IN CONTINUOUS-FLOW ANALYSERS

In Technicon AAI technology, the flash distillation option found many applications. It is a simple clean-up procedure which removes a whole range of interfering compounds from the reagent stream of interest. However, since the introduction of AAII technology, it has tended to be forgotten. To some degree, this can be explained by the lack of an acceptable commercially designed unit. The value of the technique to work at the Laboratory of the Government Chemist was obvious and followed early work by Mandl *et al.* [10] and Keay and Menage [11], and most closely, the design of a more flexible unit by Shaw and Duncombe [12]. This latter unit was used for the determination of aldehydes and ketones in experimental cultures of micro-organisms.

A device described by Sawyer and Dixon [13] was used for the determination of alcohol and acid in beer and stout. Attempts to improve the reliability of this method and to improve the signal-to-noise characteristics of the measurements prompted a critical design described by Lidzey *et al.* [14]. This unit overcomes many of the fluctuations in results observed with use of the first unit: in this a number of possible sources of surging were indicated and these were not controlled owing to the varying conditions in the coil. In addition, the separation of the waste involatile material from the volatile phase took place outside the heated flask distillation unit. Air bubbles present in the segmented stream were also responsible for considerable surging.

The improved system is shown in Fig. 4.2.

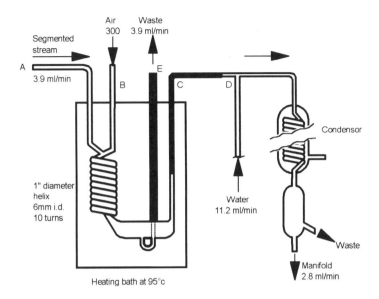

Fig. 4.2   Improved flash distillation apparatus for continuous-flow analysis.

The air-segmented sample stream enters the coil at (A) which contains none of the glass beads of the original unit to restrict the flow. At the point of entry the air bubble is released immediately and the liquid meets the carrier-gas stream (B) and flows in a thin film continuously down a coil with the air continually flowing over it; alcohol and some water vapour are vaporized into the gas stream. At the bottom of the coil, residual liquid is pumped to waste via the vertical tube (E), and the vapour flows through the capillary junction (D) where water is introduced. The gas/liquid stream then passes through a vertical condenser and condensate collects in a liquid trap with a wide exit tube; the liquid is then resampled through the manifold and analysed by the dichromate reaction. For long-term operational performance, modifications to the heating-bath temperature control are necessary and these have been described by Bunting *et al.* [15].

Use of this device for routine analysis of real-world samples again illustrates the need to understand the chemistry involved. It was used in the analysis of beer and wine for alcohol, acid and sugar content. Whilst the device performed reliably with alcohol-water solutions, the results produced with beer samples were low and erratic. This was because the sample contained certain proteinaceous materials which affected the alcohol distillation rates and hence gave variable results. However, the addition of detergent to the wash-water stream was found to overcome this fluctuation, and reliable and consistent results were then obtained. Modifications of the method for the analysis of samples of wine has also highlighted a similar problem. Wines can contain up to 30% w/v of sugar, and such a wide variation has deleterious effects on the distillation rate. However, the addition of a solution of 2% sugar and 2% ammonia to the wash-water serves to improve the distillation characteristics, swamp out the variation caused by the sugar and neutralize any acid present.

A reliable flash distillation unit as described above has many applications for AAI methodology, and it can also be used coupled to AAII technology. For the determination of sulphur dioxide in wine and soft drinks it has many advantages over the commercial methods based on a gas membrane.

Sulphur dioxide is used as a preservative in foodstuffs. General concern about the quantities of this preservative in the diet necessitates periodic monitoring in the laboratory. The distillation system described previously for the determination of alcohol in beers and wines has found further application in the determination of total sulphur dioxide in wines and soft drinks. Figure 4.3 shows a diagram of the procedure. Essentially the system is derived from the procedure developed by Morfaux and Saris [16] for wine and uses the method for fixation of the sulphur dioxide described by West and Gaoeke [17]. Samples are acidified with phosphoric acid and the sample and air are distilled in the unit described above. Sulphur dioxide distils from the solution and is fixed with a solution of potassium tetrachloromercurate. The potassium disulphate mercurate formed is reacted with formaldehyde and *p*-rosaniline producing a red solution of *p*-rosaniline methylsulphonic acid which is measured colorimetrically at 560 nm. A range of soft drinks and wines has been analysed by the automatic method described previously and the results compared to those obtained using an SO_2 electrode system and the manual method. Comparative results published by Bunton *et al.* [18] show that the automatic distillation system described compares favourably with the accepted manual method. Conventional autoanalyser (AAII) technology for the analysis of SO_2 typically employs the use of a gas dialysis membrane. The advantage of the distillation system is that it provides more efficient transfer of SO_2 and permits a lower limit of detection.

Phosphoric acid is used in place of sulphuric acid to avoid the introduction of SO_2 into the sample. In practice, the automatic method operates at the rate of 12 samples per hour to avoid inter-sample interaction and a linear response is obtained for concentrations between 0 and 15 ppm. Samples with higher levels of SO_2 can be analysed after an initial dilution with SO_2-free water. No attempt was made to provide data-processing facilities for this method. A simple computer program is used to calculate a calibration line, and, hence, the values of SO_2 in the samples.

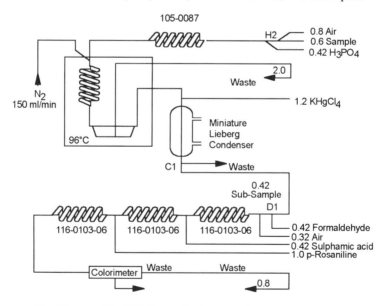

Fig. 4.3 Manifold diagram for the automatic determination of total SO_2 using a flash distillation technique.

4.4 CHROMATOGRAPHIC APPLICATIONS

Many applications of chromatography have been automated to some degree: attention has focused primarily on the automatic injection of samples or on data processing and reporting. However, the separation power of the column can be usefully exploited as a pretreatment process.

4.4.1 Hybrid gas-chromatographic systems

In an automatic instrument developed by Lidzey and Stockwell [19] for the analysis of furfuraldehyde in gas oil, a preliminary separation is performed on a GC column coupled to a specific colorimetric reagent in a continuously flowing liquid stream. A back-flushing unit is incorporated into the instrument, which extends the life of the column and removes heavy hydrocarbons from the gas stream. The method provides a single-peak chromatogram for samples containing furfuraldehyde, which is identified by: (1) a response signal, and (2) the corresponding retention time. The most important feature of the instrument is the design of the interface between the GC outlet and the flowing liquid stream. The instrument uses a conventional GC with an automated syringe injection and with the conventional detector replaced by the interface to the colorimeter. The instrument provides specific detection which can be readily extended to other analytes; some possible further applications, including the measurement of aldehydes in tobacco-smoke condensates, have been described by Rickert and Stockwell [20].

Hydrocarbon distillates in the gas oil range ('diesel' or 'derv') are subject to duty when used as a road fuel. Gas oil, which is often identical to diesel oil in hydrocarbon composition, is exempt from duty when used for stationary machines. In order to prevent its misuse as a road fuel, gas oil is marked with a mixture of 1,4-dihydroxyanthraquinone (quinizarin), 2-furfuraldehyde (furfural) and a red dye. An automatic method for the extraction, identification and determination of quinizarin in gas oil has been used by the Laboratory of the Government Chemist (LGC) for some years. The presence of furfural provides evidence for legal prosecution and the numbers of analyses ordered in the UK merit automatic analysis.

A specification for such an automatic instrument can be summarized as follows:

1. The analytical procedure should give a positive, unequivocal identification of the presence of furfural.
2. Furfural should be determined in the range 0 to 6 mg kg^{-1}, with a precision of ± 0.02 mg kg^{-1} (a fully marked gas oil contains 6 mg kg^{-1} of furfural).
3. In order to reduce the development time and to aid serviceability and maintenance, the instrument should be constructed from readily available components and modules wherever possible.
4. In order to be able to cope with the numbers of samples to be analysed annually, the maximum analysis time must not exceed 15 to 20 minutes.

The standard procedure that had been in use for a number of years at the LGC involved two separate analyses: determination by an aniline acetate colorimetric method and identification by thin-layer chromatography. Direct automation of this procedure into a single automatic method was not practicable and several other approaches to the problem were given consideration, for example modification of the solvent-extraction principles described in the quinizarin system so that the furfural could be extracted, and the use of the gas-chromatographic identification or the use of a gas-chromatographic method incorporating a trapping system between two columns. Graham [21] has described a system similar to the last-mentioned method for the determination of furfural in tobacco-smoke condensates. Both of these options were feasible, but it was considered that the identification of a component peak on the basis of a single retention time would provide insufficient evidence for the unequivocal identification required. The advantages and disadvantages of the various methods for the determination of furfural have been discussed in more detail elsewhere [22].

A simple and more direct method, which involves the use of a primary gas-chromatographic separation combined with a specific colorimetric device, was evaluated against the standard manual technique and proved sufficiently reliable to form the basic design for a routine automatic instrument.

The resulting instrument is self-standing and all components are mounted on a specially constructed movable bench. Instrument controls are mounted on the front panel of the instrument and the device requires only the supply of the following services: 240V a.c. 50Hz mains electricity, nitrogen and compressed air. The equipment consists of three integral modules: the gas chromatograph with autosampler; the scrubbing unit; and the colorimetric detection system.

A gas chromatograph is used for the primary separation of the components in gas oil. An automatic unit feeds the chromatograph with samples and a back-flushing unit has been added in order to remove heavy hydrocarbons, which might otherwise choke the column. The carrier gas (nitrogen) is controlled by three electropneumatic valves, as shown in Fig. 4.4. In the normal mode with valve A open and valves B and C closed, carrier gas flows through columns 1 and 2 to the detector.

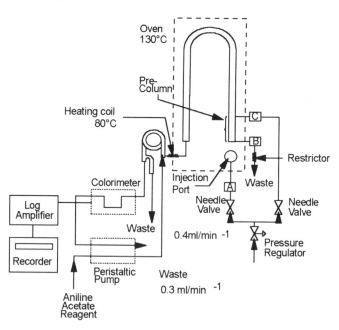

Fig. 4.4   Schematic layout of gas and liquid flow lines. A, B and C are electropneumatic valves.

With B and C open, and A closed, back-flushing purges column 1 of heavy organic compounds. The system requires only a single pressure regulator, the gas flow being controlled by needle valve restrictors. Two 6 mm diameter columns are used in the analysis, the first being 300 mm long and the second 1.5 m long, packed with Chromosorb G impregnated with Carbowax 20M. The separations are carried out isothermally at 140°C. The automatic injection device is used in the conventional manner except that an electronic timer with a modified injection controller is used in order to initiate and control the operation of the back-flushing valves. Injections of 10 µl of gas oil samples are used, these being compatible with the column configurations, so that the colorimetric method must therefore be sensitive to 0.6 mg kg^{-1} of furfural.

4.4.1.1 Scrubbing unit

The scrubbing unit is the most important part of the analytical system and optimization of its performance is essential for precise operation. The chemical reaction involved produces an unstable colour, so the residence time must be sufficient for colour development to occur but the solution must be transferred rapidly to the detector so as to minimize loss of colour intensity. The unit consists of five component parts whose particular spatial configuration has been determined experimentally: the liquid inlet from the peristaltic pump; the gas inlet from the gas-chromatographic column; the mixing coil, in which the furfural is transferred into the liquid stream and the colour developed; the debubbling unit, which separates nitrogen from the liquid; and the liquid trap from which liquid is transferred to the detector. Figure 4.5 shows the detailed configuration of the scrubbing unit.

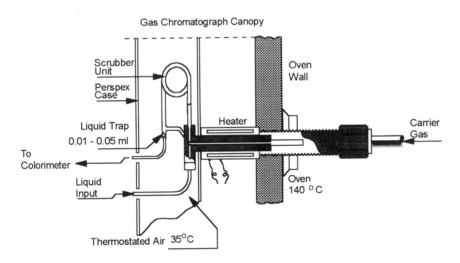

Fig. 4.5   Detail of scrubbing unit.

Carrier-gas is transferred from the column through a heated metal capillary, which minimizes the 'dead' volume at the end of the column and prevents condensation of the column effluent prior to its entry into the scrubbing unit. The tube carrying the liquid stream is joined to the gas stream tube, at a T-junction that is joined to the mixing coil by a glass-to-metal seal. Furfural is transferred from the gas stream into the liquid stream and the colour develops; the two phases are then separated by the debubbling unit and the liquid stream is re-sampled through the flow cell of the colorimeter. A Technicon peristaltic pump is used to control a differential flow-rate into and out of the unit so that the liquid trap remains full during the analytical procedure. Tygon pump tubing has been found suitable for this application, while other flow lines are constructed from PTFE tubing. Bubbles of nitrogen are almost entirely eliminated from the colorimeter by this procedure; the remaining bubbles appear as sharp spikes that cannot be confused with a chromatographic response on the chart record.

The complete scrubbing unit is controlled in an air thermostat, constructed from Perspex, at $35° \pm 1°C$. In order to give a quick response to changes in the ambient temperature, the thermostat control incorporates a solid-state proportional controller and a fan that circulates air over a bare-element heater.

4.4.1.2 Colorimeter

It is important to minimize the lengths of the transfer lines to the colorimeter flow cell and to use readily available components in order to reduce development time. In order to achieve these criteria, the optical section of the colorimeter was separated from the electronic controls and placed immediately adjacent to the scrubbing unit, within the housing of the gas chromatograph. The electronic controls are mounted on the front panel of the instrument below the gas chromatograph. Liquid from the scrubbing unit is pumped through the flow cell (path length 20 mm and internal volume 200 ml) fitted with a 520 nm filter. The normal colorimeter response is modified, by using a logarithmic amplifier, to give an output that is linear with concentration, and recorded on a strip-chart recorder.

The particular configuration described above produced extremely well-defined chromatographic peaks, without any gross peak broadening effects being evident.

Injection of 10 ml of gas oil containing furfural in the range 0-10 mg kg^{-1} gives a chromatogram with a single peak. A retention time of 9.5 min for the peak includes the transit time through the scrubber to the optical cell. A negative spike on the recorder trace marks the injection point; no solvent peak is observed with gas oil in this detection system. A typical recorder trace is shown in Fig. 4.6. A graph of the concentration of furfural in gas oil against peak height was found to be linear. However, drift in sensitivity occurs over a period of several hours owing to decrease in reagent sensitivity and variation of flow-rates at the peristaltic pump.

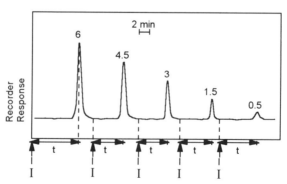

Fig. 4.6 A typical recorder trace showing a range of standard injections in the range 0.5 - 6.0 mg kg^{-1} of furfural. I, injection point; and t, retention time.

In order to compensate for these variations, a standard solution of furfural is injected every fifth test. The furfural concentrations in samples are calculated by calculating the mean peak height of the standard solution before and after the sample set, and using a linear interpolation.

The automated system based on these principles gives positive identification and quantification of furfuraldehyde in a complex matrix. It has been vigorously tested for reliability and performance. Clearly, the technique could be applied to other problems.

4.4.2 Automatic analytical gas-chromatographic systems

The majority of commercial developments which relate to the automation of GC and HPLC pay little attention to sample preparation. There are few examples where pre-treatment is not required. A fully automated system was developed by Stockwell and Sawyer [23] for the determination of the ethanol content of tinctures and essences to estimate the tax payable on them. An instrument was designed and patented which coupled the sample pre-treatment modules, based on conventional AutoAnalyzer modules, to a GC incorporating data-processing facilities. A unique sample-injection interface is used to transfer samples from the manifold onto the GC column. The pretreated samples are directed to the interface vessel by a simple bi-directional valve. An aliquot (of the order of 1 ml) can then be injected on to the GC column through the capillary tube using a time-over pressure system.

A schematic diagram of the automatic system is shown in Fig. 4.7a. The sample-to-gas-chromatograph interface and requirements for the injection of a liquid mixture onto a gas-chromatographic column from a flowing stream present the greatest problem, in contrast to sample transfer and dilution techniques in automatic analysis, which are well documented.

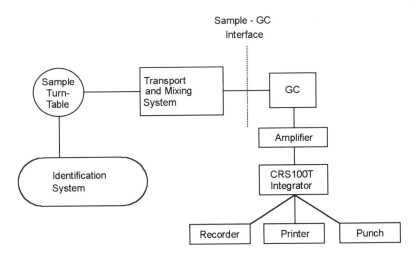

Fig. 4.7a   Schematic diagram of automatic system.

In order to satisfy the specification outlined above, a special interface was developed. The detailed design of this interface, the transport system, and over-pressure system have been the subject of a previous publication and a patent application [24]. The system devised for the interface is based on modifications to an injection system presented by Heilbronner *et al.* [25] for preparative gas chromatography. A commercial version of the preparative unit was developed by Boer [26]. In this application, samples are monitored into the injection vessel which is connected to the gas chromatographic column by a length of stainless steel capillary tubing.

The injection is performed by forcing a slug of sample mixture through the capillary onto the column by an overpressure of carrier gas. The carrier-gas inlet, capillary inlet, and the top of the column form a 'T' junction. For optimum performance this unit should have a minimum dead-space and should also be robust. A suitable unit (shown in Fig. 4.7b), can be constructed by a straight Simplifix coupling, which has a gas inlet tube brazed into the side wall. The capillary and the column are positioned so that carrier gas leaks into the column system through the joint between the column and capillary. Both the column and the capillary have carefully machined ends, a single scratch is made on one surface with a file, and this allows adequate throughput of carrier gas. The injection vessel is constructed from Pyrex glass and has a capacity of 10 - 20 ml. Its size and overall configuration have been optimized experimentally.

In the normal rest position, carrier gas is fed to the top of the column at pressure P1 through a gauge G1 and a buffer volume B1. Excess of carrier-gas bleeds back through the injection capillary, cleaning out the tube, and is allowed to vent to atmosphere via the side arm (d), which is connected through the buffer B2 and the valves V2 and V3.

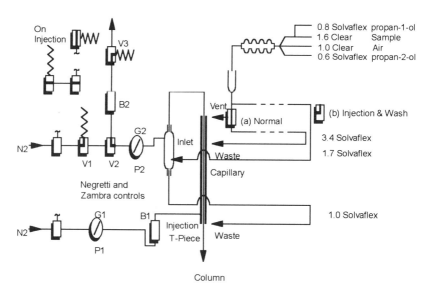

Fig. 4.7b   Schematic of suitable unit.

The gas at pressure P2 is normally vented to waste via the valve V(1) and restrictor R1. To produce an over-pressure injection, valves V(1), V(2), and V(3) are activated and the over-pressure is applied to the injection vessel, thereby forcing liquid through the capillary on to the column. Pressure in the vent line between V(2) and V(3) is released simultaneously, via the buffer B2, to the atmosphere. This operation allows for the rapid release of surplus gas pressure and the minimum disturbance to the gas chromatographic conditions. The volume of B2 was determined by experiments at the operating pressures. The quantity of liquid injected is determined by the Hagen-Poiseuille Law and the injection is dependent on the viscosity of the sample mixtures and the over-pressure that is applied. The samples in this application covered a range of preparations containing ethanol, and, in order to obtain comparable injections for each solution, the sample is first swamped with sufficient internal standard solution to minimize the variation of viscosity. The diameter of the injection capillary was chosen to allow injections compatible with the size of the column used.

The object of the transfer system is to dilute the sample and provide adequate mixing to form a representative mixture for the gas chromatography injection vessel.

Samples and standard solutions are placed in test tubes and sealed with Parafilm caps, the tubes are loaded onto the turntable, and, during a complete set of operations, the machine executes the following functions:

1. The tubes are located in sequence at the probe position and during the 'tube change' operation a simple flag system identifies whether a tube contains a sample or a standard; the probe is then lowered into the tube.

2. The sample is pumped, together with air and internal standard, through the mixing unit.

3. When a uniform mixture is obtained, the injection vessel is filled, drained, and re-filled.

4. The sample mixture is injected by application of an over-pressure.

5. The probe is raised and re-positioned over the vessel containing a supply of the internal standard solution.

6. The probe is lowered into the internal standard solution and a wash-out cycle is performed.

7. The probe is raised and re-positioned over the turntable.

8. At the end of chromatogram, the turntable finds the next tube and a new cycle is started.

Data-processing facilities can be provided by using either commonly available IBM PC compatible software or hardware integration.

4.4.3 Automatic liquid-chromatographic systems

Burns [27] described a fully automated approach for HPLC analysis of vitamin tablets. A sample valve provides the injection interface in this application. Tablets direct from the production plant are dispensed into the sample cups on a Technicon 'Solid Prep' sampler, they are dissolved, and the fat-soluble vitamins are extracted. The solution is concentrated by a novel evaporation-to-dryness module which removes most of the solvent, and the vitamins are re-dissolved in the appropriate volume of solvent. The analysis is performed by injecting a suitable sample of this stream through a sample loop of an injection valve onto a liquid-chromatographic column fitted with a conventional liquid chromatographic UV detector. A schematic diagram which illustrates the system is shown in Fig. 4.8. It can be extended to gas-chromatographic analysis by interfacing the sample-preparation system to the gas-chromatographic column.

Another approach to sample pretreatment has been described by Hoogeveen *et al.* [28] for the determination of the organochlorine pesticides in milk and milk products.

The milk is deproteinized and extracted with hexane by a novel continuous extractor. After on-line filtering and drying, the extract is injected onto a liquid chromatograph consisting of a precolumn and an analytical column. The precolumn provides a sample clean-up which retains the fat and separates the organochlorine compounds into two classes. Sampling, pretreatment, clean up, chromatographic separation, and detection and column regeneration are all performed under microcomputer control.

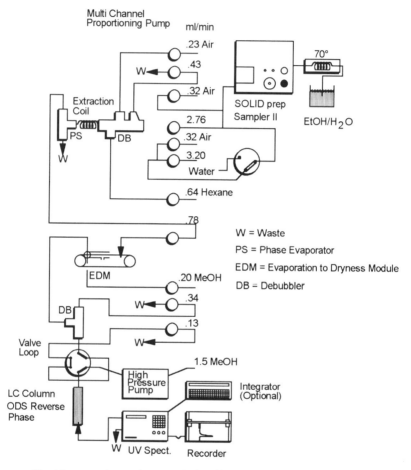

Fig. 4.8   Automatic system designed by Burns [27] for the analysis of vitamin tablets. *Reproduced with permission of the Technicon Instrument Co. Ltd.*

4.4.4   Sample-preparation for biochromatography

In the separation of biomolecules, sample preparation almost always involves the use of one or more pretreatment techniques. With high-performance liquid chromatography (HPLC), no one sample preparation technique can be applied to all biological samples. Several techiques may be used to prepare the sample for injection. For example, complex samples require some form of prefractionation before analysis, samples that are too dilute for detection require concentration before analysis, samples in an inappropriate or incompatible solvent require buffer exchange before analysis, and samples that contain particulates require filtration before injection into the analytical instrument.

The sample-preparation technique may depend on a number of variables, for example the molecular weight of sample and interferences, the sample volume and analyte concentration, buffer salt (anion and cation) content and metal concentration and type. Other than filtration for particulate removal, most of the approaches are based on the use of chromatographic media for cleaning up samples before analysis.

Two approaches are considered: the batch technique and the column technique. In the former, the stationary phase (medium) is 'dumped' into the sample in a liquid form, usually in an aqueous environment. The stationary phase is allowed to contact the sample for a period of time and then is removed by filtration or by pouring off the liquid phase, leaving behind the compound of interest sorbed onto the stationary phase or contained in the liquid phase. The batch technique is slower than the column approach, but it is very easy to perform.

The column approach - in which the sample is passed through the medium - is more widely used. Although dilution is a possible consequence, the column technique is more useful for removing traces of the analyte of interest. Convenient prepacked sample preparation columns are readily available. A newer form of medium is the flow-through membrane disk, which offers less flow resistance than a column yet provides similar sample capacity. Majors and Hardy have provided a thorough review of the range of methods in use [29].

4.4.5 Automatic preparative gas chromatograph

Figure 4.9 shows the schematic layout of a fully automatic, commercially available, gas chromatograph. This has been designed for fast effluent separation and isolation of pure fractions at rates from as little as a few mg to 125 g per hour.

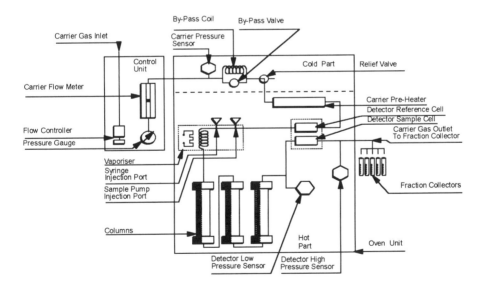

Fig. 4.9   Schematic layout of a fully automatic gas chromatograph.

Samples are injected into the vaporizer by a metering pump or manually with septum injection; the manual injection procedure is intended for method development. The sample gas mixture then passes through the chromatographic column where the sample compounds separate. Fractions pass through the thermal conductivity detector and then to a condenser collection manifold where up to five fractions can be collected. Complete control of the system is achieved via a mini-computer.

Automatic operation of the metering pump allows repetitive injections with unattended operation. Precise control of the carrier-gas flow ensures stable chromatograms and reproducible timings for collection of samples. Independent column oven and vaporizer temperatures are available up to 300°C and these can be operated with temperature programming or isothermally. The latter option is the most common.

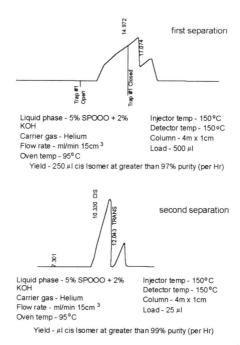

Fig. 4.10 Chromatograms showing the cis Isomer separation of 2,6-Dimethylmorpholine on a Gas Chromatograph.

The detector has been specifically designed for stable operation. Figure 4.10 shows a schematic representation of an analytical application. The sample is separated twice; on the first pass the fraction collected is indicated on the chromatogram. This fraction is then re-injected and the first peak is collected.

The system has been used for a wide range of applications, the secret of its success being the attention to detail and a series of interlocking safety alarms. Several key parameters are continuously monitored and the built-in safety alarms will automatically shut the system down if there is a fault. As can be seen from Fig. 4.10, the balance the user has to make with these systems is between column overload and throughput. As the size of the injection increases, so the quality of the chromatogram decreases and the purity of the fraction suffers. A compromise has to be reached in terms of the purity desired and the throughput that has to be maintained.

4.4.6 Automation to achieve full potential of gas chromatography

In 1995 Hewlett Packard introduced the HP 6890 series of instruments which are designed to fulfil a broad range of analytical requirements. This series includes a specific automatic liquid sampler for increased throughput and automation, a series of options for data-handling/control for the gas chromatograph and samples, as well as an integrator that supports the full range of features provided.

The system shown in Fig. 4.11 is designed to deliver optimal performance when coupled to sample-preparation, sample-introduction and data-handling products available. The HP 6890 series GC offers a smooth transition for users of HP 5890 series gas chromatographs through methods compatibility, extremely useful for modern laboratories whose methodology is costly to develop.

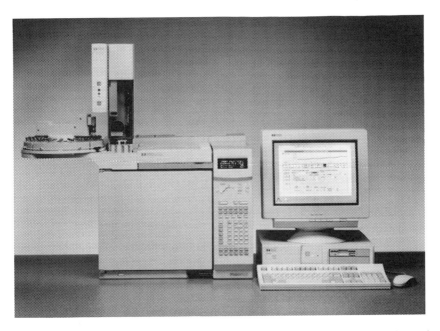

Fig. 4.11   The HP6890 series GC system. *Reproduced with permission of Hewlett Packard Company.*

The HP 6890 series features electronic pneumatics control which provides complete electronic control of all gas pressures and flows. On-board sensors automatically compensate for ambient temperature changes and barometric pressure differences to provide more accurate and reproducible results routinely. This reduces recalibration frequency and improves laboratory productivity. It also decreases system operating costs, allows a faster set-up time, and reduces equilibration time after set-points are changed from one method to the other.

The pneumatics also optimize the performance of the split/splitless inlet. Its forward pressure control in splitless injection mode significantly reduces the risk of sample loss and maximizes accuracy and reproducibility. The provision of mass flow control coupled with back pressure control in the split injection mode maximizes reproducibility and accuracy and also allows electronic adjustment of split ratios. The net result is that sample-introduction conditions are optimized individually for the two most popular injection techniques. In addition, parameters are recorded in the methods file.

The new GC has been designed for maximum ease-of-use and gives chromatographers the ability to achieve accurate, reproducible results from run to run and from user to user.

The ease-of-use features include the following:

1. All pneumatic parameters, including split ratios, can be set from the keyboard to eliminate routine use of the bubble meter.
2. An expanded colour-coded keyboard gives analysts direct access to control functions without the need for shift keys.
3. A five-method storage capability for all GC set points reduces set-up time and increases reproducibility.
4. An information key defines all set points and advises analysts on appropriate input ranges to make methods set-up fast and easy.
5. A four-line alphanumeric display guides analysts through methods set-up with step-by-step prompting for set-point parameters.
6. Built-in HP-IB and RS-232 communication ports provide easy system interconnection and allow the GC to operate in diverse laboratory environments.

The HP 6890 series also offers a range of automation features to improve laboratory productivity, increase system uptime and reduce the cost of ownership.

Automation features include the following:

1. Clock-time programming loads and runs a method for complete unattended operation.
2. Pre-run programming enters conditions and prepares the system for the next injection.
3. Gas saver reduces split flow rate after the start of a run and between runs to conserve expensive carrier-gas.
4. Post-run programming starts after the chromatographic run, stops data acquisition, and sets pressures, flows and temperatures to clean out the system.

Built-in diagnostics capabilities, such as a run log for status information and modular pneumatic manifolds, inlets and detectors, reduce system maintenance and improve uptime.

The HP 6890 series GC gas-sampling valve technology, in conjunction with programmed interval sampling, makes this type of GC-based quality-control automation a reality. The GC can control and sense the position of multi-position valves to make it possible to automate sample-stream selection. Each sample, which is downloaded automatically, can then be analysed by a different method. Automated analysis of multiple process streams is especially attractive and cost-effective for complex chemical and petrochemical operations in which raw materials for product streams need to be assessed at regular intervals.

Another system component, the automatic liquid-sampler system, offers faster tray operations and improves vial handling to increase system throughput. The liquid-sampler system can be controlled through the GC keyboard or the HP ChemStation.

Capabilities of the ChemStation include a full-featured standard reporting package and a custom report designer for analysts with special reporting requirements. An optional database software package provides standard cross-sample and study reports as well as trend analysis. The HP ChemStation also improves data transfer to other software, and networking facilities are available to link into a corporate computing system.

The key features and benefits of the multi-technique ChemStation include the following:

1. Multi-technique data-handling: Laboratories can improve productivity and save on training costs by using the same software package to acquire and analyse data from their HP, GC, LC, CE and A/D systems.

2. Automation: Users can fully automate their systems from pre- and post-run programming to multi-method sequencing to meet specific analysis requirements.

3. Good Laboratory Practice (GLP): The system provides a comprehensive feature set to aid customers in meeting GLP requirements. This includes features such as certificate-of-software validation, user-access levels, instrument and sequence logbooks, system-suitability software for all supported HP instruments, standard GLP reports, and a GLP save option that encrypts and saves data and methods together.

4. Flexible reporting capability: The ChemStation software provides a versatile, standard reporting package. Those users with customer requirements can use the system's custom report generator or can use Dynamic Data Exchange to export the data to other Microsoft packages for reporting.

5. ChemStore database: This optional software package is a powerful extension of ChemStation software for sample organization and results storage. Control charts and cross-sample and cross-study reports are standard and enhance the ChemStation's already versatile reporting software.

6. Industry standards: Only standard formats are used to enhance usability and productivity with other laboratory programmes.

7. Networking: For laboratories with multiple ChemStations, networking facilities are provided for greater productivity and enhanced regulatory compliance.

A new integrator offers full compatibility and support of the HP 6890 series GC features, and includes gas-saver mode, split valve control and post-run programming. It also contains two new industry-standard Personal Computer Memory Card International

Association (PCMCIA) slots to accommodate long-term data storage and prints the run-deviation log after each analysis.

With HP 6890 series GC electronic pneumatic control, all GC parameters, including gas control, are recorded automatically for every run to improve adherence to Good Laboratory Practice (GLP) and to ease regulatory compliance. This is of particular importance in the petrochemical industry. A run-time log records all changes in set-points during a run.

In addition, the HP ChemStation provides password protection against accidental data loss or methods changes. The HP ChemStation controls and monitors all GC parameters and maintains a logbook of all system events that occur while the GC system is running. System-suitability software, which allows analysts to select from a wide variety of chromatographic parameters to monitor and verify system performance, is also available.

For analysts in the pharmaceutical industry or other regulated industries, data and methods are stored together in a binary, read-only file to ensure adherence to regulatory requirements. Improvements in the design of detectors have also been integrated into the HP 6890 to improve ruggedness in users laboratories. Through automation the benefits of the separation technologies are optimized.

4.5 AUTOMATION OF SAMPLE PREPARATION OF FOODSTUFFS FOR TRACE METAL ANALYSIS

Widening interest in the quality of the environment has led to increased demand for information on a wide range of trace-metal contents of foodstuffs. Trace metals in foodstuffs are normally determined by spectroscopic techniques after complete destruction of the organic matrix. Destruction is achieved either by wet oxidation or by dry ashing; additional treatment is normally required in order to obtain the metals of interest in a form suitable for analysis. Both methods of destruction are time consuming and tedious; this is particularly true of the wet-oxidation procedure, which has the additional disadvantage of being potentially hazardous; the methods require considerable analytical skill and experience. Both methods are prone to produce erroneous results either by the loss of an element of interest or by adventitious contamination from the component parts of the apparatus used. Stockwell and Knapp [30] have described the available models and these will be reviewed later.

At the Laboratory of the Government Chemist, increased analytical demands and the expense of suitable staff resulted in a requirement for the complete, or partial, mechanization or automation of this kind of work. A review of the literature, together with an examination of the procedures then used in the laboratory, suggested that a wet-oxidation system would be the most suitable for automation. This is described in detail here because it provides a valuable lesson in systems design. The specification of the automatic system was that it should be modular in construction in order to facilitate modification and maintenance, and that it should have a multi-element capability.

Jackson et al. [31] have constructed and evaluated a system based on the above design criteria. The design of their system, the method of operation and the performance of the digester system are also described here.

4.5.1 Apparatus

A schematic diagram of the automatic system is shown in Fig. 4.12. It consists of five component modules: a sample introduction unit, a digestion unit, a neutralization vessel, a chelation and extraction vessel and an extract collection unit. The system is controlled by a series of interacting cam and electronic process timers.

The method of sample preparation, the design of the sample introduction module and the general problem of transferring samples and acids from one module to another were the subject of considerable investigation. Two particular problems placed serious constraints on the types of materials used in the construction of the modules: the risk of physical damage to the system due to the extremely corrosive chemicals used (concentrated sulphuric and nitric acids and ammonia), and the possibility of trace elements from the components being introduced into the system and producing false results. These problems were largely overcome by the choice of inert non-corrodible materials, for example, Kel-F, or by plating components with a noble metal such as gold.

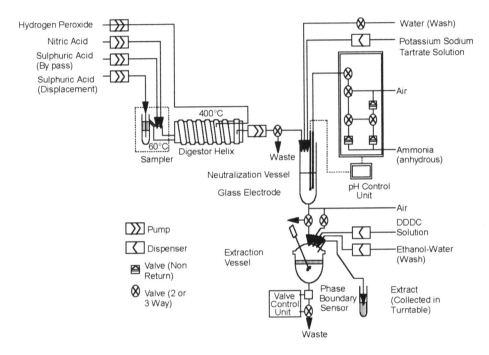

Fig. 4.12 Schematic diagram of the system for automatic sample preparation of foods for trace metal analysis.

4.5.2 Sample introduction unit

The method of sample preparation and the design of the sample introduction module were the result of a considerable amount of preliminary investigation to overcome a number of problems. The system must be capable of dealing with a range of samples with very different chemical and physical properties. Contamination of the sample by metals during the preparation stage must be avoided. The digester must be fed with sample and acids at constant rates and for a specified period. The whole of the sample or sub-sample must be transferred to the digester without leaving a residue. As it is both difficult and dangerous to make accurate volumetric measurements of foodstuffs suspended in 33% sulphuric acid, these measurements should be avoided.

These problems were solved in the following way. A manual method of sample preparation was chosen to allow maximum flexibility in the treatment of a range of samples. Whenever possible, the samples were freeze-dried and powdered. To minimize the possibility of contamination, samples were prepared for analysis in polyethylene bags. Quantitative measurements were performed by weighing the sample and bag at various stages in the preparative procedure set out below, with an automatic balance and data-recording system comprising a Mettler PT 1200 top-loading balance connected to an ASR-33 teletype through a Mettler CT-10 interface module. This balance has a capacity of 1200 g and a precision of ± 10 mg. The teletype produces a paper-tape record of the indicated mass on receipt of a command initiated by a push-button on the interface unit.

The sample introduction unit was constructed from inert materials, which minimizes the introduction of metal contamination into the system. The samples or digestion acids make contact only with PTFE, Kel-F, glass, acid-resistant rubber and platinum-iridium (9 + 1) alloy. In addition, the construction materials were limited to acid-grade Arborite (ureaformaldehyde laminate), Perspex and stainless-steel. The unit was constructed in three continuous sections: a heated sample compartment, a turntable mechanism and heat-exchanger compartment, and a pump compartment.

4.5.3   Heated sample compartment

The heated sample compartment is maintained at 60°C and contains an Arborite turntable that holds 20 glass sample displacement vessels with side-arms. A pneumatic actuator is connected to a pair of platinum-iridium probes that can be lowered to the bottom of the sample tubes or raised to clear the lid of the compartment. A funnel-shaped mixing chamber, made from Kel-F and PTFE, is located below the sample tube side-arm, the tail of the funnel passing through the well of the sample compartment and into the digester helix. Two ports above the funnel allow the introduction of streams of concentrated sulphuric and nitric acids. A reed switch, fitted to the wall of the sample compartment, is actuated by a small permanent magnet that can be fixed to the turntable near the last tube in the batch, thus shutting down the system and leaving it in a 'stand-by' condition. Transparent Perspex top and front panels are fitted to the sample compartment to allow inspection during operation; these panels can be removed to allow insertion and removal of the turntable. The turntable base is sealed and fitted with a rim large enough to contain acid in the event of a spillage.

4.5.4   Turntable mechanism and heat exchanger

The turntable mechanism and the heat exchanger are situated below the sample compartment. An Arborite disc is fixed securely to a central stainless-steel spindle on which the turntable is also mounted. A horizontal pneumatic actuator is fitted beneath this disc in such a way that, when it is extended, it pushes against one of a set of 20 pegs mounted on the under-surface of the disc and rotates the turntable. A second pneumatic piston mounted vertically below the disc can be extended to force a conical pin into one of a set of 20 holes drilled in the underside of the disc. This locking device has two functions: it positively locates the side-arm of the sample vessel over the funnel, and it prevents the turntable from being moved accidentally. Electrically interlocking switches provide an additional safeguard by preventing the locking and rotating actions being carried out simultaneously and by inhibiting rotation when the probes are lowered.

The heat exchanger consists of a small centrifugal fan that takes air from outside the unit and forces it, via ducts, into the four corners of the sample chamber. A heater made from Nichrome wire wound around a ceramic former is situated within the ducting immediately upstream of the fan. The temperature within the sample chamber is monitored by a thermistor set in the floor of the chamber and a proportional temperature controller maintains the temperature at $60^\circ C$.

4.5.5 Pump compartment

In the pump compartment, situated at the base of the unit, separate pumps are provided for the displacement sulphuric acid, the by-pass sulphuric acid, hydrogen peroxide and the removal of waste displacement acid from the sample tubes. The displacement acid is dispensed by a purpose-built syringe pump; single-channel peristaltic pumps are used for the other liquids.

The syringe pump, apart from the stepping motor and control circuitry, is constructed from corrosion-resistant material; the acid comes into contact only with glass, PTFE and Kel-F. The syringe is driven by a stepping motor, which permits discharge at a constant and reproducible rate of 2.0 ml min^{-1} and refilling at 6.0 ml min^{-1}, coupled with fast manual operation in either direction. The limits of syringe travel, and therefore the volume dispensed per stroke, are determined by a rod of suitable length that is attached to the piston mounting and which moves between two microswitches. Operation of either of these microswitches stops the syringe drive and also switches a pneumatically actuated three-way valve, fitted at the end of the syringe, to either fill or discharge the syringe.

The by-pass sulphuric acid, the nitric acid and the hydrogen peroxide streams are pumped by peristaltic pumps which are fitted with Acidflex tubing on the acid streams and Tygon tubing on the hydrogen peroxide stream. These pump tubes are connected by modified couplings to PTFE tubing, which is used for most of the plumbing in the sampler unit. The flow-rates are monitored with all-glass flow meters. A peristaltic pump is used to remove waste displacement acid.

4.5.6 Digestion unit

The digestion unit consists of a modified Technicon continuous digester with ancillary equipment for introduction of samples, removal of the digest and disposal of fumes. The continuous digester consists of a borosilicate-glass helix, which is rotated over three banks of heaters; the samples and digestion acids are fed in at one end, transported over the heaters by the rotation of the helix, and the resulting digest is pumped out of the other end of the helix for subsequent analysis. A long glass tube, inserted at the output end of the helix, allows the addition of hydrogen peroxide to complete the digestion process.

The control of the temperature of the standard digester was found to be inadequate, the desired operating temperature being attained by adjusting the current to the heaters so that it balanced heat losses. Fitting proportional heating controllers to each of the three banks of heaters resulted in greater flexibility in the operation of the digester, a more rapid attainment of the working temperature, and the virtual elimination of temperature drift. A description of these modifications, and the resulting improved performance of the digester, has been published previously [32].

The output of the digester is pumped from the helix through a pneumatically actuated valve either into the neutralization vessel or to waste. A pump was designed and built for this purpose. Its action is based on an oscillating piston, it has no moving valves and the hot acid comes into contact only with Kel-F and PTFE. The maximum flow-rate is about 12 ml min^{-1}. This pump has been described in more detail elsewhere [33].

4.5.7 Neutralization unit

This unit consists of a jacketed glass vessel with an Arborite lid and a motor-driven three-way glass stopcock with PTFE key fitted to the outlet. Fitted into the lid are a demountable, refillable combination glass pH electrode and inlet tubes for digest, a dilution (buffer solution, the gaseous ammonia - air neutralization mixture) and the wash water. The jacket is maintained at $25^{\circ}C$ by a bath of circulating water; this temperature was chosen to give adequate cooling without the risk of ammonium sulphate crystals forming, because these could block the outlet valve.

The pH signal is fed via a pH meter to a potentiometric recorder and to a neutralization controller, which compares the electrode e.m.f. with pre-selected values and opens or closes gas-control valves accordingly. Depending on the pH, it is possible to obtain a flow of air, ammonia or a mixture of both gases. By careful adjustment of the potentiometers and the flow-rates of ammonia and air, it is possible to control the final pH and also to keep the digest within a fairly closely defined pH range during the neutralization process.

Efficient and continuous mixing is essential for proper control of the neutralization process. This is achieved by introducing a stream of air into the stem of the vessel immediately above the outlet valve. It is not sufficient to rely on mixing by the incoming stream of neutralizing gas because when pure ammonia is being used the gas dissolves so readily that no turbulence is induced and a dense layer of acid tends to collect at the bottom of the neutralization vessel.

An automatic dispenser is used to add a 2% m/v solution of potassium sodium tartrate to the neutralization vessel before the digest is introduced. This protects the electrode and helps to reduce the concentration of dissolved salts. Two jets mounted in the lid provide a spray of water to wash the walls of the vessel, the electrode and the gas inlet tube. Washing is initially carried out with the outlet valve closed and a vigorous stream of air passing through the inlet tube; the valve is then opened to drain the vessel and washing is continued with the outlet valve opened. A commercial version of this automatic netralization stage was introduced in 1991 by the Knapp Logistic company in Austria.

4.5.8 Chelation and extraction unit

This unit is positioned directly below the neutralization unit and is constructed from a modified glass reaction vessel fitted with a standard reaction head with inlets for sample, water wash, ethanol - water wash, chelating reagent solution, compressed air stirrer and extract removal probe. An outlet in the bottom of the vessel is connected to a phase boundary detector, which is connected in turn to a valve that can be switched to a waste line or to a collection vessel. Automatic dispensers are used to add both the chelating reagent (a 2% m/v solution of diethylammonium diethyldithiocarbamate [DDDC] in heptan-2-one) and the ethanol-water wash liquid. The latter is used to remove trace amounts of residual organic solvent and it is therefore stirred vigorously, allowing it to splash off the top of the extraction vessel and run down the walls. This liquid is run off before the vessel is finally rinsed with water, which is admitted through a perforated circular glass tube fitted within the reaction vessel head.

The extraction vessel outlet valve can be controlled either manually from the control unit or automatically by the phase boundary detector unit. The operation of this detector is based on the difference in refractive index between the organic and aqueous phases, and has been described elsewhere [34].

The organic phase is removed by applying a positive air pressure to the extraction vessel, thus forcing the extract up a PTFE tube and through a probe into a glass, or polypropylene, extract collection vessel situated in the extract collection module. During the washing cycle, the tube and probe are washed out with ethanol - water and then blown dry with air.

4.5.9 Extract collection unit

The extract collection unit consists of a turntable with provision for holding 20 glass or polypropylene tubes. Two pneumatic pistons are used to raise and lower the probe and to rotate it through $90°$ to position it either over a tube for collection of extract or over a waste outlet for rinsing during the wash cycle. Both the rotation of the turntable and the movement of the probe are controlled by the central control unit.

4.5.10 Control system

The control system is largely situated in a series of 19" rack-mounted units. As far as possible, each of these units is self-contained with one controlling the sample introduction unit, one the neutralization unit, one the chelation - extraction and extract collection units and another being used as the source of the various power supplies. The neutralization controller and phase boundary sensor are situated in small separate units and the digestion unit controller is incorporated into the digester module.

The system control is based on a series of six cam timers, which control the following: sample introduction unit, sampling phase; sample introduction unit, refilling phase; neutralization unit, neutralization phase; neutralization unit, wash phase; chelation - extraction unit, extraction phase; and chelation - extraction unit, collection and wash phase.

4.5.11 Performance

The performance of the system was clearly demonstrated for a wide range of foodstuffs. The data for the NBS (National Bureau of Standards) bovine liver (Table 4.1) shows that the automatic system is capable of giving accurate results. Samples were freeze-dried on receipt. Measurements on manually digested samples were made by atomic-absorption spectrophotometry, and by plasma-emission spectrometry on automatically digested samples. The agreement between automatic and manual determinations was very satisfactory.

Table 4.1   Analysis of NBS standard reference material 1577 (bovine liver)

Element	NBS value*, µg/g	Measured†, µg/g
Fe	270 ± 20	262 ± 7
Cu	193 ± 10	189 ± 7
Zn	130 ± 10	133 ± 6

*   The NBS values are based on the results of 6-12 determinations by at least two methods.
†   The values from this work are the means of nine determinations.

4.6 RECENT DEVELOPMENTS IN AUTOMATIC SAMPLE PREPARATION

4.6.1 Trace elemental analysis

Significant advances have been made over the last few years to improve the performance and detection limits of trace elemental analytical techniques. The VG Elemental and Perkin Elmer (Sciex) ICP Mass Spectrometers, for example, offer detection levels down to a few parts per trillion for many elements. However, it is still necessary to be able to introduce the samples effectively into these instruments. This, however, puts increasing pressure on the analyst to devise effective sample preparation techniques which will retain the advantages gained by using the instruments. What is required is to establish rules and guidelines for developing a strategy for reliable trace analysis in the ng/g or pg/g region. Knapp [35] has a good reputation in this area and has established criteria which are listed in Section 4.6.1.2.

4.6.1.2 Basic rules for trace analysis (Knapp [35])

1. All materials used for apparatus and tools must be pure and as inert as possible. Generally this requirement is met by quartz, Teflon, glassy carbon, and, to a lesser degree, by polypropylene.

2. The apparatus and containers must be scrupulously cleaned to provide low blank levels and to lower losses due to absorption or the ingress of contamination released from the vessel walls.

3. To minimize systematic errors, micro-chemical techniques should be used with small apparatus with optimal surface to volume ratio. All procedures should be carried out in a single vessel, and, if volatile materials are to be measured (for example, selenium), a closed vessel must be used.

4. The reagents, carrier gases and auxiliary materials must be as pure as possible.

5. Good housekeeping is essential so that high blanks are not caused by dust or material contaminating the reaction vessel.

6. The analytical procedure should be kept as simple as possible. The best sample preparation process is, in fact, none.

7. All stages of the analytical process *must* be effectively monitored and tested by comparison with peer laboratories.

These rules may appear very daunting but there are a number of tools commercially available for trace analysis. Before defining the best analytical procedure, it is important to define clearly the objectives of the analytical procedure. The analytical procedure and the instrumental requirements must be considered: how many elements need to be determined, what level of precision is required and in what form are the samples submitted for analysis? Another important parameter is the time available to obtain the result.

Table 4.2 gives a summary of methods for sample decomposition [36-40]. Fluxes are particularly important for inorganic materials. The use of fluxes is limited by the reaction of trace elements within the crucibles and by the high blank levels in the fluxing materials themselves.

Table 4.2   Summary of decomposition methods

DECOMPOSITION METHOD

Fusion Decomposition

Wet Chemical Decomposition	(a) in open systems
	(b) in closed systems (e.g. bombs)
Combustion	(a) in open systems
	- dry ashing
	- combustion in a stream of oxygen
	- oxidation with excited oxygen
	(cool plasma ashing)
	(b) in closed systems
	- oxygen flask (e.g. Schoniger flask)
	- combustion in oxygen bombs
	- combustion in a dynamic system
	(Trace-O-Mat)
	(c) the Wickbold combustion method

Dissolution of biological materials with tetra-alkylammonium hydroxide (e.g. Lumatom)

For the decomposition of organic materials, wet chemical methods are most frequently used. The wide acceptance of these methods is due to the multitude of decomposition reagents available, simple handling procedures and high sample throughput. The error sources listed above can largely be eliminated if wet chemical methods are properly handled.

For the oxidation of organic materials with air or pure oxygen a variety of techniques has been established which produce very good results. Dry ashing, however, although widely used in the past, should be completely abandoned from the protocols of trace analysis because the experimental parameters are poorly reproducible.

4.6.2   Development of better decomposition methods

Wet digestions in open vessels may or may not involve refluxing. Since it is critical to adhere closely to the optimized decomposition parameters of time and temperature, mechanization of the decomposition not only leads to higher sample throughput with less human intervention but also to the avoidance of errors. The simplest form of mechanization can be implemented through a time- and temperature-controlled heating block. A greater degree of mechanization would also incorporate control of reagent reflux during decomposition [41].

These procedures operate batch-wise. Continuous sample handling has some advantages over discontinuous handling. The samples are not all available at the same time and can be analysed as soon as they complete the digestion step. Therefore, no major time lag occurs between decomposition and analysis in which element losses and contamination could take place. Moreover, continuous systems generally match the analytical needs better than batch procedures. For example, the Automatic Wet Digestion VAO Anton Paar (Graz, Austria) system is such a continuously operating digestion system. The samples are fed in from one side and passed through the digestion zone [42].

Table 4.3 Wet chemical decomposition in open vessels

	Microwave Decomposition (Prolabo Microdigest)	**Automated Wet Decomposition Device VAO**
Sample	organic and inorganic	organic and inorganic
Amount	up to 2 g	up to 2 g
Decomposition time	3/45 min	30/180 min
Throughput	2/20 samples/hour	10/40 samples/hour
Advantages	• high flexibility with automated version (type A 300)	• continuous operation allows constant time intervals between end of decomposition and measurement (to prevent losses of Hg and I)
	• system for reagents addition available	• different glassware and PTFE vessels available - simply convertible
		• proven design for continuous operation without supervision
		• maximum temperature of 400°C
Disadvantages	• batch-wise operation	
	• expensive (automated system)	
Remarks		ideal instrument for labs with high throughput of similar samples. All methods of wet chemical decomposition can be performed

Wet decomposition in open vessels, however, does not reflect the latest achievements in this field. A relatively large amount of reagent leads to elevated blanks, so that tackling samples in the ng and sub-ng range is rarely possible.

Two arguments can be put forward in favour of wet chemical decomposition methods in open vessels, firstly, price, and, secondly, sample throughput. A wet decomposition can, if necessary, be performed with a beaker and a heating plate, whereas apparatus for modern high performance decompositions can cost as much as $25 000. With respect to sample throughput, the automatic wet digestion device can be used for up to 150 samples per hour for some materials. Such throughput cannot be realized by any other method. Some special cases exist that necessitate decomposition in open vessels (for instance, when the sample needs to be treated with perchloric acid).

Recently, Prolabo (12 Rue Pelee, 75011 Paris, France) has introduced a fully automated open vessel digestion system using a focused microwave source as the heating assembly. This system has a fully programmable sequencer and acid dispensing system and is a cost-effective method of automating such open vessel digestions. Table 4.3 compares and contrasts the use of microwave decomposition with the Anton Paar VAO design.

4.6.3 Decomposition methods in closed vessels

The state of the art for the decomposition of biological materials are the ashing methods. Of course, some of these methods still have teething troubles. The most recent include the following techniques and devices:

1. Wet chemical decomposition in a closed system, with nitric acid or mixed acids consisting of nitric acid, hydrochloric acid, perchloric acid, hydrogen peroxide and hydrofluoric acid. The following devices are available:

    Decomposition in closed Teflon vessels at medium pressure (up to 8 bar) with microwave heating [43-44]. Microwave Digestion System MDS-81D (CEM Corporation, North Carolina, USA; Floyd Inc., South Carolina, USA).

    Decomposition in closed Teflon vessels at high pressure (up to 85 bar) with microwave heating. (Microwave Acid Digestion Bomb, Parr Instrument Company, USA.)

    Decomposition in closed quartz or glassy carbon vessels at high pressure (up to 120 bar) with conventional heating [45-47] High Pressure Asher (Anton Paar Company, Graz, Austria).

2. Low temperature ashing with radio frequency induced oxygen plasma:

    Cool Plasma Asher CPA 4 (Anton Paar Company, Graz, Austria).

3. Combustion in a dynamic system with complete recovery of all volatile elements:

    Trace-O-Mat VAE-II (Anton Paar Company, Austria).

The classical PTFE decomposition bomb is not discussed here because it is no longer state of the art. If the well-known disadvantages of PTFE as vessel material are accepted, a combination with microwave heating will be advantageous.

Table 4.4 compares the most important information about the three decomposition systems. The main differences lie in the method of heating and the vessel material. The decomposition methods of the CEM Corporation, Floyd Corporation and Paar

Instrument Company systems apply microwave heating, whereas the High Pressure Asher has a conventional heating system.

Table 4.4   Wet chemical decomposition in closed vessels

	PTFE - Bombs (various manufacturers)	Microwave Decomposition in PFA Pressure Vessels	High Pressure Asher HPA (Pressure decomposition in quartz vessels)
Sample	organic and inorganic	organic and inorganic	organic and inorganic
Amount	up to 2 g	up to 3 g	up to 1.5 g
Maximum temperature	200°C	180°C (locally higher temperature in the solution)	320°C
Decomposition time	60 -120 mins	3 - 30 mins	60 - 150 mins
Throughput	depends on the number of bombs used	up to 50 samples/hr	3 - 7 samples/hr
Advantages	• cheap, when only one bomb is used	• short decomposition time • high sample throughput • easy to use (CEM, Floyd)	• quartz vessels allow extreme • high oxidation potential of nitric acid at 300°C destroys all organic substances - no interference with voltametric or hydride systems
Disadvantages	• PTFE (or PFA, FEP) vessels limit the max. temperature to 200°C - organic residues may remain • other problems with vessel material: adsorption, diffusion leaching, memory effects • safety risks	• venting of the vessels (losses of elements) because of too fast reactions when reaction progress cannot be foreseen • temperature limitation - incomplete decomposition	• limited sample throughput
Remarks	useful, when occassionally single samples have to be decomposed	for large numbers of similar samples, where low accuracy is sufficent	for versatile use, especially where high precision is required (ultratrace levels)

Microwave heating allows the decomposition time to be decreased by a factor between 3 and 10. Microwave heating has advantages for inorganic samples in terms of reduced decomposition times. For decomposing silicate minerals only one to five minutes are needed, whereas 1 to 2 hours are necessary when applying a High Pressure Asher. For organic samples ranging from food to plastics, 15 - 60 minutes are necessary when applying microwave decomposition and 60 - 120 minutes when using a High Pressure Asher. Teflon is used as vessel material for the microwave decomposition systems, whereas High Pressure Asher vessels, made of pure quartz glass, are suitable for all acids except hydrofluoric. For decomposition with hydrofluoric acid in the High Pressure Asher, glassy carbon is an option.

Comparison of the fields in which these three decomposition methods are used leads to the following picture: for the analysis of inorganic sample material, microwave systems with their very short decomposition times are advantageous. Element concentration is comparatively high in this field and systematic errors from adsorption or diffusion into the vessel material can be observed when using Teflon are not significant. In the analysis of biological samples, which very often have an extremely low element content, significant systematic errors may occur with Teflon decomposition vessels, and the High Pressure Asher, with its decomposition vessels made of highly pure and inert quartz glass, is advantageous in this area. A further benefit of the High Pressure Asher is the complete destruction of all organic compounds with nitric acid at 320°C [48]. At 200°C, the maximum temperature of microwave ashing with Teflon vessels, not all organic compounds are destroyed [49]. This leads to systematic errors in the determinations of arsenic and selenium by Hydride AAS or Hydride ICP OES [50] and with voltametric determinations of heavy metals. The slightly longer decomposition time should be accepted in favour of the lower error rate.

In 1989 Prolabo introduced the A300, the first truly automated microwave digestion system. The A-300 incorporates the focusing and reflux design of the first Microdigest system, a carousel for sequential loading and digesting of up to 16 samples, and a reagent-adding system that permits automated addition of up to four reagents, including HF. This system controls both the volume and rate of reagent dispensing. The latest configuration of the system is shown in Fig. 4.13.

Fig. 4.13   The Prolabo A- 301 microwave digestion system. *Reproduced with permission of Prolabo.*

Other features include storage and recall of digestion procedures, whereby up to four procedures can be stored in the A-300's computer. In addition, an unlimited number of procedures may be stored and recalled from a PC-type computer through the A-300's RS-232 interface. The system permits the operator to program each vessel independently. When a batch of 16 samples is loaded, there is no need for all of them to be equal. The first six samples may call for an HNO_3 procedure, the next four may require an HF addition, the next five may require an H_2O_2, and the last sample may even be an H_2SO_4 and H_2O_2 digestion. All of these operations can, if needed, be treated as one batch. The operator merely weighs in the sample and loads the carousel.

Low-frequency plasma ashing and the other techniques such as the Trace-O-Mat products have been developed at the University of Graz, Austria. Commercial developments outside Austria and Germany have been slow owing to the need for manual involvement. Further details on these products can be found in the publication by Knapp *et al.* [51].

4.6.4 Applications

Except for some very difficult materials that require very high pressure [52], the Microdigest systems can be used for virtually any type of sample and could be described as a 'thoroughly modern hotplate'.

Some of the key areas of application for focused open-vessel microwave digestion have been described by Grillo [53].

4.7 ALTERNATIVE APPROACHES

Two new ways of automating sample preparation have been commercially exploited; firstly the infra-red reflectance techniques, which avoid much of the sample pretreatment required for conventional analysis, and, secondly, robotics, to fully automate or mechanize manual techniques.

Near infra-red reflectance spectroscopy was first used by Norris and Hart [54] for the determination of moisture, oil and fat in cereal products. A number of instrument companies, notably Technicon, have developed commercial instruments. The instrument replaces a series of chemical procedures by a signal measurement in each of six infra-red regions and reference to a suitable computer calibration. Such an approach offers considerable advantages. Although the instrumentation available has been developed for the analysis of cereals, it has wider applications, for example in the tobacco industry. It does, however, have some drawbacks. The instrument must be calibrated against a suitably accurate standard method, and unfortunately few chemical methods are available. Also, the six wavelengths were selected for cereal analysis, and are not necessarily suitable for other materials. On the other hand, one of the major advantages of the technique is that the analysis can be done on site, away from the laboratory.

The Technicon InfraAnalyser range has been designed to meet these requirements and is a good example of how, by the introduction of current technology, a flexible instrument results. The incorporation of microprocessor technology allows self-calibration and self-teaching aids. The computer system is also fully expandable to cater for a range of applications. Applications in wheat, dairy, animal feed, tobacco and cocoa analysis and many other areas are under evaluation. Some of the major obstacles to the application of the technique to a wider range of samples, especially the inflexibility, have been overcome by this instrument.

The Zymate laboratory automation system combines robotics and laboratory stations to automate procedures used in sample preparation. A micro-processor-based controller acts as an interface among operator, robot and laboratory station, managing the sample preparation procedures, their sequence and their timing. Under control of this unit the robot transfers samples form station to station, according to user-programmed procedures. For example, when dispensing, the robot moves a test-tube or vial containing the sample to the dispenser to introduce the programmed amount of reagent or solvent. The controller waits for a signal from the dispenser that the operation is complete and then instructs the robot to move the sample to the next operation. When the sample preparation is complete, the robotic arm either introduces samples directly into the analytical instrument or places them in a carousel or rack for subsequent analysis.

The basic system includes the controller with user memory, robot, a general-purpose hand, and the capacity for six laboratory stations. These approaches will find further use as they are applied to varying sample types. Further details of robotic systems are discussed in Chapter 6. The continuing series of automation conferences organized by the Zymark Corporation provide a ready access to the latest advances in robotic technology.

Both of the above approaches have limitations, the first due to difficulties in relating results to calibration data, and the second because it becomes prohibitively expensive to cater for every situation in an instrumental approach. However, in this respect Zymark has made significant progress by the introduction of the Benchmate products. These are tailored to a specific market area such as the sample preparation of pharmaceutical products prior to HPLC analyses. Thus, cost-effective solutions can be swiftly tailored to specific market areas. As progress is made on these products, it is likely that third party vendors will start to generate specific modules for individual customer needs.

REFERENCES

[1] Carter, J.M. and Nickless, G., *Analyst*, 1970, 95, 148.
[2] Coverly, S.C. and Macrae, R., *Journal of Micronutrient Analysis*, 1989, 5, 15.
[3] Karlberg, B. and Thelander, S., *Analytica Chimica Acta*, 1978, 98, 1.
[4] Vallis, G.G., UK Patent Application 14964/67, 1967.
[5] Anderson, N.G., *American Journal of Clinical Pathology*, 1970, 53, 778.
[6] Bartels, H., Werder, R.D., Schurmann, W. and Arndt, R.W., *Journal of Automatic Chemistry*, 1978, 1, 28.
[7] Williams, J.G., Stockwell, P.B., Holmes, M. and Porter, D.G., *Journal of Automatic Chemistry*, 1981, 3, 81.
[8] Trowell, F., *Laboratory Practice*, 18, 144, 1969.
[9] Stockwell, P.B., *Proceedings of Analytical Division of Chemical Society* 1975, 12 (10), 273.
[10] Mandl, R.H., Weinstein, L.H., Jacobson, J.S., McCure, D.C. and Hitchcode, A.E., *Proceedings of Technicon Symposium 1965*, Mediad Inc., New York, 1966, 270.
[11] Keay, J. and Menage, P.M.A., *Analyst*, 1970, 95, 379.
[12] Shaw, W.H.C. and Duncombe, R.E., *Proceedings Technicon Symposium 1966*, Mediad Inc., New York, 1967, 75.
[13] Sawyer, R. and Dixon, E.J., *Analyst*, 1968, 93, 669.

[14] Lidzey, R.G., Sawyer, R. and Stockwell, P.B., *Laboratory Practice*, 1971, 20, 213.
[15] Bunting, W., Morley, F., Telford, I. and Stockwell, P.B., *Laboratory Practice*, 1976, 23 (4), 179.
[16] Morfaux, J.N. and Saris, J., *Connaissance de la Vingre et du Vin*, 1971, 4, 505.
[17] West, P.W. and Gaoeke, G.D., *Analytical Chemistry*, 1956, 28, 1816.
[18] Bunton, N.G., Crosby, N.T., Jennings, N. and Alliston, T.G., *Journal of Association of Public Analysts*, 1978, 16, 59.
[19] Lidzey, R.G. and Stockwell, P.B., *Analyst*, 1974, 99, 749.
[20] Rickert, W.S. and Stockwell, P.B., *Journal of Automatic Chemistry*, 1979, 1, 152.
[21] Graham, J.F., *Bietrage zur Tabakforschung*, 1969, 5, 43.
[22] Stockwell, P.B., *Laboratory Practice*, 1978, 27, 715.
[23] Stockwell, P.B. and Sawyer, R., *Analytical Chemistry*, 1970, 42, 1136.
[24] Stockwell, P.B. and Sawyer, R., UK Patent Application 10550, 1969.
[25] Heilbronner, E.E., Korals, E. and Simon, W., *Helvetica Chimica Acta*, 1957, 40, 2410.
[26] Boer, H., *Journal of Scientific Instrumentation*, 1964, 41, 365.
[27] Burns, D.A., *Proceedings of the Seventh Technicon International Congress*, New York, 1977.
[28] Hoogeveen, L.P.J., Wilmott, F.W and Dolphin, R.J., *Analytical Chemistry*, 1976, 282, 401.
[29] Majors, R.E. and Hardy, D., *LC GC International*, 1992, 5 (6), 10.
[30] Stockwell, P.B. and Knapp, G., *International Labmate*, 1989, 14 (5), 47.
[31] Jackson, C.J., Porter, D.G., Dennis, A. and Stockwell, P.B., *Analyst*, 1978, 103, 317.
[32] Jackson, C.J., Morley, F. and Porter, D.G., *Laboratory Practice*, 1975, 24, 23.
[33] Lidzey, R.G., Jackson, C.J. and Porter D.G., *Laboratory Practice*, 1977, 26, 400.
[34] Porter, D.G., Jackson, C.J. and Bunting, W., *Laboratory Practice*, 1974, 23, 111.
[35] Knapp, G., *ICP Newsletter*, 1984, 10, 91.
[36] Bock, R., *A Handbook of Decomposition Methods in Analytical Chemistry*, International Textbook Company Ltd, London, 1979.
[37] Henschler, D., *Analysen in Biologischem Material*, Leseblattausgabe, 1976, 2, 23.
[38] Kelker, H., *Ullmanns Encyklopadie der Technischem Chemie*, 1980, 5, 27.
[39] Gorsuch, T.T., *The Destruction of Organic Matter*, Pergamon Press, Oxford, 1968.
[40] Dolezal, J., Povondra, P. and Sulcek, Z., *Decomposition Techniques in Inorganic Analysis*, Iliffe Books, London, 1968.
[41] Seiler, H., *Laborpraxis*, 1979, 6, 23.
[42] Knapp, G., Sadjadi, B. and Spitzy H., *Fresenius Z. Analytical Chemistry*, 1975, 274, 275.
[43] Knapp, G., *Aerztl Laboratory*, 1982, 28, 179.
[44] Kingston, H.M. and Jassie, L.B., *Analytical Chemistry*, 1986, 58, 2534.
[45] Knapp, G., *Trends in Analytical Chemistry*, 1984, 3, 182.
[46] Knapp, G. and Grillo, A., *American Laboratory*, 1986, 3, 76.

[47] Scharmel, P., Hasse, S. and Knapp, G., *Fresenius Z. Analytical Chemistry*, 1987, 326, 142.
[48] Wurfels, M., Jackwerth, E. and Stoeppler, M., *Fresenius Z. Analytical Chemistry*, 1987, 329, 459.
[49] Welz, B., Schubert-Jacobs, M., Schlemmer, G. and Sperling, M., Lecture at 5th International Workshop Trace Elemental Chemistry in Medicine and Biology, 1988, Neuherberg, Germany.
[50] Kaiser, G., Tschopel, P. and Tolg, G., *Fresenius Z. Analytical Chemistry*, 1983, 316, 482.
[51] Knapp, G., Raptis, S.E., Kaiser, G. and Tolg, G., *Fresenius Z. Analytical Chemistry*, 1981, 308, 97.
[52] Knapp, G. and Grillo, A.G., *American Laboratory*, 1986, 18 (3), 76.
[53] Grillo, A., *Spectroscopy*, 1989, 4 (7), 16.
[54] Norris, K.H. and Hart J.R., *Proceedings of International Symposium (1963) on Humidity and Moisture*, Reinhold, New York, 1965.

5

Automated atomic spectroscopy

5.1 INTRODUCTION

The major goals for the future development of analytical atomic spectrometry measurements are improved detection limits and the development of simple ways of coupling to other analytical techniques. The nebulizer systems of the spectrometric instruments are the parts that need to be improved in order to achieve these goals. Typically, nebulizer efficiencies are of the order of 1-2%, and, as a result, they are limiting factors for instruments which can cost between £100,000 and £150,000.

Backstrom and co-workers [1] have demonstrated significant improvements in nebulizer efficiencies by increasing analyte transport efficiency, at a solvent load acceptable for the atom reservoir in question, and therefore improving detection limits. Conventional nebulizer systems do not allow this because an increased analyte transport efficiency will give a too high a solvent load in the atom reservoir.

The nebulizer system can be regarded as an interface between the sample and the instrument [1-4]. Gustavsson [2] described an interface which uses a jet separator for decreasing the amount of solvent vapour introduced into the atom reservoir. The interface gives an analyte transport efficiency of approximately 35%. It is intended for the coupling of low flow-rate techniques, for example miniaturized flow-injection analysis (FIA) and microbore liquid chromatography (LC), to the inductively coupled plasma (ICP), the microwave-induced plasma (MIP), or graphite-furnace atomic-absorption spectrometry (GFAAS).

An analyte transport efficiency of nearly 100% has been obtained with an interface for flame atomic absorption spectrometry (FAAS) [3]. It has been used for the determination of lead in blood [5] and for coupling with a high-performance liquid chromatograph (HPLC) [6].

The on-line coupling of analytical techniques to the ICP has been chiefly limited to techniques using aqueous solutions. Organic solvents are a major problem which can be minimized by using a membrane interface [4], rather than a conventional nebulizer system. Gustavsson [4] described the characterization of the interface using chloroform and Freon as solvents. There are, however, applications demanding more polar solvents.

5.1.1 Membrane interface

The membrane interface coupled to an ICP is shown schematically in Fig. 5.1. Two separators have been studied, with membrane thicknesses of 12 and 15 μm respectively. The interface with the thicker membranes used a Meinhard TR-20-C2 nebulizer rather than the more common TR-20-B2, because it is cheaper and better suited for this application (less cavitation). The outlets of the separators were at room temperature (20°C).

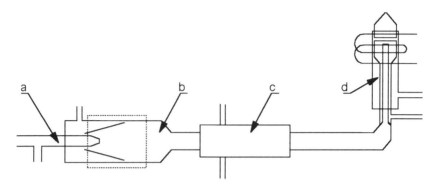

Fig. 5.1   Membrane interface designed by Gustavsson for ICP [4].

The interface consists of: (a) the nebulizer; (b) the chamber; (c) the membrane separator and (d) the torch. Organic solutions are fed to the nebulizer via a peristaltic pump and the aerosol chamber is heated by a 250 W heating tape around the chamber. The membranes used are made from silicone rubber and this is supported by high density polyethylene supports. A vacuum pump provides the vacuum on the membrane to remove the solvent vapour.

The alternative jet separator concept was originally developed to couple gas and liquid chromatographic systems to mass spectrometers. The separator modified for atomic spectroscopy consists of an evacuation vessel with two concentric tubes having inner diameters (inlet/outlet) of 0.25 mm respectively and a spacing of 1 mm. The gas and solvent vapour in the aerosol are sucked off to vacuum, whilst the aerosol passes across the gap between the tubes and is transferred into the detector via the plasma torch. Organic solvents and water can both be removed in this way. Gustavsson [7] has described both of these devices in considerable detail. Significant improvements were made to detection levels for a wide range of elements, often by an order of magnitude. The extension to organic solvents, especially polar solvents, has also been exploited.

Table 5.1 shows a typical set of results illustrating the improvements made by using this type of separator.

The procedure for measuring the solvent removal efficiencies has been described elsewhere [4]. All of the detection limits in the work described here were estimated as the concentration giving a signal equal to three times the standard deviation of the signal at the analytical wavelength when aspirating a blank. The lowest value reported for the results of the determinations is take to equal twice the detection limit.

For the ICP-CFE-D and ICP-CFE-I techniques, the continuous-flow extraction system was coupled directly or indirectly to the membrane interface. Using the indirect technique, the extract was collected in a volumetric flask capped with aluminium foil and subsequently aspirated into the membrane interface. The determinations were performed by standard additions, and mean values were based on 10 replicate measurements. The ICP instrument was also used in conjunction with the membrane interface without preconcentration; the detection limits, in this case, were based on 10 replicate measurements.

Table 5.1 Comparison of detection limits obtained for the ICP system designed by Gustavsson with those for ICP atomic-emission spectrometry (AES) and ICP-MS according to Montaser and Golightly [8], in comparison with those obtained for ICP-CFE-D and ICP-CFE-I

Method	\multicolumn{6}{c}{Detection limit (ng ml^{-1})}					
	Cu	Co	Fe	Mn	Ni	Zn
ICP-AES	3.2-9.9	3.2-9.9	3.2-9.9	1.0-3.1	10-31	1.0-3.1
ICP*	1.6	1.8	1.8	0.3	4.2	0.54
ICP-CFE-D	0.12	0.17	0.092	0.30	0.45	0.27
	(3.9)	(2.9)	(2.4)	(3.0)	(5.7)	(5.7)
ICP-CFE-D†	0.12	0.11	0.063	0.35	0.90	
	(2.1)	(1.2)	(0.9)	(4.2)	(12.6)	
ICP-CFE-I	0.024	0.057	0.032	0.048	0.23	0.23
	(0.44)	(0.78)	(0.58)	(0.53)	(3.1)	(3.7)
ICP-MS	0.030	0.020	0.20	0.04	0.03	0.08

* Water uptake rate of 1.8 ml min^{-1}.
† Results obtained, excluding zinc, during the optimization.

The values in parentheses represent the RSDs of the signals in percentages for a concentration of 3 ng ml^{-1}.

Note: ICP-CFE-D = Inductively Coupled Plasma-Continuous Flow Extraction-Direct coupling to membrane interface
ICP-CFE-I = Inductively Coupled Plasma-Continuous Flow Extraction-Indirect coupling to membrane interface

5.2 HYDRIDE/VAPOUR-GENERATION TECHNIQUES

Vapour-generation techniques have been successfully designed to improve the analytical performance for many of the trace elements causing environmental concern. Such techniques have been developed specifically for the Group IV and V metalloids. Arsenic, selenium, antimony, bismuth, tin, tellurium, germanium and lead have been determined as the gaseous hydrides. Mercury, which exists as atoms in the gas phase at room temperature, can be determined as a vapour. With normal nebulization sample introduction techniques, only a small percentage of the sample actually enters the measurement detector and the sensitivity is rather poor. Consequently the levels of these elements in the environment cannot be detected. Hydride/vapour-generation techniques ensure that 100% of the element is presented to the measurement device, as well as providing increased sensitivity and selectivity.

Results presented by Stockwell [9] for some of the hydride-forming elements and for mercury illustrate the enormous increase in sensitivity achieved with automated analytical chemistry methods (Table 5.2). Earlier developments centred on the batch approach. These methods have recently been dropped (in favour of continuous-flow techniques) because they were not easy to use, were very dependent on operator ability, and were difficult to automate.

Table 5.2 Comparison of detection levels for normal nebulization and vapour generation on plasma 2 ICP spectrometer using a PSA 10.003 Hydride Generator

Element	Detection limits (µg/l)			
	Expected performance	Actual performance	Performance with vapour generator	Improvement factor
As	100	430	0.7	614
Se	200	320	0.7	457
Sb	70	130	0.2	650
Te	60	80	0.3	267
Bi	20	50	0.1	500

Table 5.3 sets out the advantages and disadvantages of the batch and continuous flow techniques. The introduction of continuous-flow hydride/vapour-generation has substantially advanced the value and acceptance of the technique for trace elemental analysis. Applied Research Laboratories (now part of Fisons Elemental), P.S. Analytical and Varian have all introduced continuous-flow hydride/vapour-generation systems, whilst Perkin Elmer has used the flow injection modification to automate the techniques with their instrumentation.

An automatic system using the hydride generation technique requires attention to (a) chemistry; (b) automatic sample preparation; and (c) introduction of the hydride into the analytical measuring device. Goulden and Brooksbank [10] designed an automatic system using an aluminium slurry, hydrochloric acid reaction scheme to convert selenium into its hydride. Other workers have successfully used the reaction with zinc and hydrochloric acid.

Table 5.3 The advantages and disadvantages of continuous and batch systems

	Continuous flow	Batch analysis
Advantages	Precise control over reaction conditions	Small sample requirement
	Constant generation of hydrogen	Economical reagent usage
	Experienced operators not required	Inexpensive equipment
	Precisions of approx. 1 % easily obtainable in linear range	
Disadvantages	Large sample volume required	Operator intensive
	Long analysis time (60 s)	Precision depends on injection technique
		Intermittent production of hydrogen
		Time consuming

However, a commercial instrument must have a broad appeal and the chemistry regime based on acid/sodium tetrahydroborate offers the most attractive approach. It has a rapid reaction, caters for all of the hydride-forming elements and can be very easily automated. To optimize the procedures, the use of standard Technicon AutoAnalyzer methodologies, i.e. matching blanks, standards and sample matrices, overcomes the difficulties due to matrix interference. Also, the detection levels achieved by a well-designed system allow the samples to be diluted, avoiding any problems still present due to matrix interference. Stockwell [9] has described a system which achieves detection levels that are an order of magnitude better than other systems.

The literature is full of papers describing interference effects of elements on the hydride forming elements, but almost all of these interferences can be overcome by a simple dilution. However, a commercial design has to cope with a broad range of applications and is not necessarily optimized either for sample types or for particular elements.

Figure 5.2 shows a schematic diagram of the automatic hydride/vapour-generator system designed by P.S. Analytical. This has been widely used to determine hydride-forming elements, notably arsenic, selenium, bismuth, tellurium and antimony, in a wide range of sample types. To provide a wide range of analyses on a number of matrices the chemistry must be very well defined and consistent. Goulden and Brooksbank's automated continuous-flow system for the determination of selenium in waste water was improved by Dennis and Porter to lower the detection levels and increase relative precision [10,11]. The system described by Stockwell [9] has been specifically developed in a commercial environment using the experience outlined by Dennis and Porter.

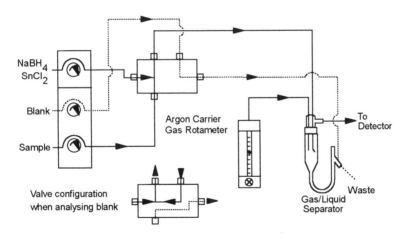

Fig. 5.2   Schematic diagram of a continuous-flow vapour generator.

To provide a consistent generation of the hydride, it is essential to have consistent pumping with minimal pulsation. No air must be introduced into the sample/reagent line and the hydride must be rapidly removed from the liquid stream. After generation of the gaseous products, an effective method of separation from the liquid stream is required. For each particular instrument measurement technique, specific transfer requirements have been optimized. Figure 5.3 shows the transfer systems required for atomic absorption, direct-current plasma and inductively coupled plasma systems. The hydrides are relatively unstable and so must be moved quickly into the measurement device. The improvements in detection levels achieved by using these techniques compared with normal nebulization are quite considerable.

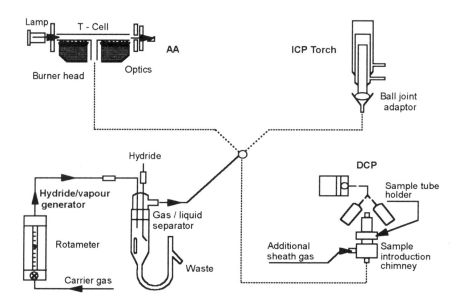

Fig. 5.3   Transfer systems to link gas-liquid separator to vapour elemental techniques.

Figure 5.4 shows a calibration graph of arsenic concentrations obtained by using a Perkin Elmer 2100 atomic-absorption system linked to a P.S. Analytical hydride/vapour generator (PSA 10.003). An electrically heated tube has been used in this work and the spectral source was an electrodeless discharge lamp. Alternatively, a flame-heated tube can be used.

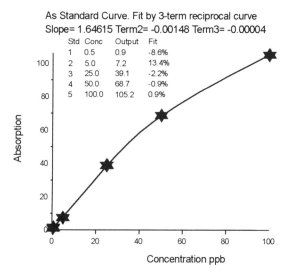

Fig. 5.4   Calibration graph for vapour generation linked to atomic absorption spectroscopy.

Typically, a reasonable signal can be measured for 0.05 ppb of arsenic in simple samples. The PSA 10.003 can be linked directly into the PE 2100 to provide a remote read signal, allowing calibration graphs and sample results to be computed. Simple interfaces have also been designed to link the system with the Perkin Elmer AS51 autosamplers, as well as to almost all other commercially available atomic absorption systems, notably Varian and Thermo Jarrell Ash. An IBM-compatible computer system, using PSA TouchStone© software, can be used to fully automate existing AA systems.

For atomic absorption the key is to transfer the sample into the T cell as quickly as possible and then to allow the hydride to remain within the cell for long enough to atomize efficiently, i.e. the hydride should remain in the T cell for sufficient time to allow atomization and for absorption to take place. Typically, the gas-liquid separator would be purged at 300-400 ml/min. Faster flows will reduce the signal, owing to the flow speed and dilution. Lower flows will reduce the signal owing to decomposition effect on the hydride.

Significant advantages can also be provided by linking the generator to inductively coupled plasma systems, particularly simultaneous systems. Conditions can be set for up to five elements. Table 5.2 shows the typical signal increases and detection level improvements. The main objective is to introduce the samples directly into the torch assembly, and particularly to puncture the plasma effectively. Typically, a flow of argon in the region of 1 - 1.5 l/min will give the optimum performance. However, it is necessary to optimize conditions for the particular type of plasma and torch configuration in use.

Introduction into a DC plasma requires rather more care and attention owing to its inherent design features. As the hydride is being introduced into the plasma, it is necessary to provide a controlled sheath of argon to contain the hydride and direct it into the plasma. This chimney effect significantly improves the sensitivity for hydride-forming elements. This interface has also formed the basis of an introduction system for mercury vapour into an atomic-fluorescence spectrometer as described by Godden and Stockwell [12].

Gaseous sample introduction into an ICP-MS presents different problems. Owing to its extremely sensitive nature, Dean *et al.* [13] introduced the sample as the gaseous hydride by a flow-injection approach. This was reasonably effective because lower volumes of samples and reagents were in use. They utilized nitric acid as a carrier stream to prevent the formation of argon chloride species in the plasma. Argon chloride has the same mass as arsenic which is mono-isotopic, and this severely limits arsenic determination. An additional problem was that the sensitivity was extremely dependent on the purity of reagents.

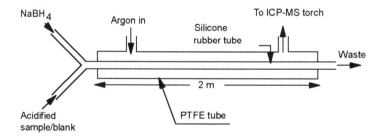

Fig. 5.5 Membrane separator designed to improve performance of ICP/MS for arsenic and selenium by vapour generation techniques.

Branch *et al.* [14] effectively eliminated argon chloride interference by using a tubular membrane gas-liquid separator (Fig. 5.5) originally developed by Cave and Green [15]. The membrane in this case allows the passage of gaseous species to the plasma, but acts as a barrier to the chloride ions. Hitchen and co-workers [16] have assessed the analytical performance of the device with ICP-MS, which is now commercially available from Fisons Instruments (Winsford, Cheshire, UK, formerly VG Elemental). Although the membrane separator does not offer 100% separation efficiency, the polyatomic interferences from argon chloride are eliminated, resulting in better performance.

5.2.1 Systems for mercury determination

Similar analytical principles can be used for the determination of trace levels of mercury. However, the sodium tetrahydroborate chemistry is often limited by the production of excess hydrogen and high levels of water vapour. The water retains the mercury, preventing it from entering the measurement system and reducing the sensitivity of the measurements. Consequently, for mercury measurements, the chemistry regime based on tin(II) chloride and hydrochloric acid is preferred. However, since this reaction does not inherently provide agitation, the gas transfer system has to be redesigned. The reaction cell is sparged with a forced flow of argon to drive out the mercury and introduce it into the measurement device. The introduction system is somewhat similar to that designed for DCP introduction and has been described by Thompson and Reynolds [17]. Figure 5.6 shows the system within the configuration of the fluorescence detector.

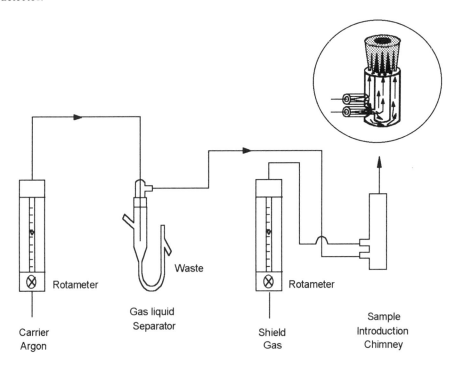

Fig. 5.6 Gas sample introduction system for mercury determination by atomic fluorescence.

The most common analytical techniques available for mercury measurements are:

1. Colorimetric.
2. Atomic absorption coupled to cold vapour generation.
3. Atomic absorption following trapping on gold from cold vapour generation.
4. Atomic fluorescence coupled to cold vapour generation.

Of these, the fluorescence approach described by Thompson and Reynolds [17] and modified by Thompson and Godden [18] has proved to be the most sensitive. Godden and Stockwell [12] have described a specific fluorescence system tailored for mercury determination in more detail. P.S. Analytical has developed a fully automated system from these initial developments to measure mercury produced from the samples, to quantify the data and provide precise analytical reports, using an IBM compatible computer and custom software.

As an addition to the automatic system, a concentration device based on amalgamation of the mercury vapour produced with gold has been developed. A vapour generator evolves the mercury which adsorbs onto a gold trap (gold gauze or gold shavings are used). This unit has been linked to atomic-absorption spectrometers. After the mercury has been concentrated for a set time, the gold is reheated to flush the mercury from the trap and a transient peak emerges at the detector. This approach significantly improves the normal absorption measurements, but adds a further complication, in that an additional device is included in the line so the analytical timescale is longer and sample throughput is reduced. The amalgamation technique has been linked directly into a number of instrumental schemes by using various vapour generators as well as a range of atomic-absorption spectrometers. The combination of preconcentration coupled to atomic fluorescence offers the lower detection levels of below 1 ppt. Cossa [19] has modified the commercially available equipment to optimize the measurement of mercury at sub ppt levels. This is described in more detail in Chapter 7.

For low-level mercury measurements, the sensitivity of the fluorescence spectrometer offers the most attractive route for analysis; it is also possible to analyse air samples. In the latter application the timescale is reduced from several hours to a few minutes. An automated system with absolute calibration techniques has been described by Ebdon *et al.* [20] and its application to air monitoring techniques set out in detail by Stockwell *et al.* [21].

Hydride/vapour generation techniques provide extremely good sensitivity. When coupled to continuous flow methodologies for use in routine analysis, simple and reliable analytical techniques are provided. The extension of chemistries and sample transfer systems to provide analytical protocols to cope with a wider range of elemental analyses should be pursued in the search for lower detection levels. While multi-element techniques offer very low levels of detection, the use of specific single element analytical instruments with detection capabilities similar to those described above may be the best route for routine laboratories with high sample throughput.

5.3 FLOW-INJECTION AND PRECONCENTRATION APPLICATIONS

A number of applications of flow-injection techniques have been made to flame atomic absorption spectrometry [22]. Although manifolds can be connected directly to the nebulizer, the response of the spectrometer is dependent on the flow rate of the sample into the nebulizer [23], and some adjustment to the manifold may be required. The optimum flow rate for maximum response when the sample enters the nebulizer as a discrete sample plug can be different from that found for analysis of a continuous sample stream.

Many of the applications described in the literature do little more than reduce the volume of sample taken. However, Perkin Elmer introduced a product particularly designed to use the flow-injection approach. At present this seems little more than a marketing ploy to gain advantages over competitors. However, in reality, many ICP systems, and indeed some AA systems, do not make it easy to link to FIA systems because the data-processing side of these instruments cannot cope with a transient rather than a continuous signal.

Where speed of response is important, or if there are difficulties with a matrix interference effect, the flow injection approach can be an advantage. Corns et al. [24] described the use of a flow injection manifold to extend the capabilities of an automated system (shown in Fig. 5.7) to determine mercury at low levels. Two applications have been researched for this approach. The first involved the analysis of concentrated sulphuric acid. Any attempt to dilute this acid would generate a great deal of heat, resulting in a possible loss of mercury. Digesting the sample in a bromate/bromide reagent, for example, would also cause problems. However, simply injecting a discrete sample of the acid into the flowing reagent stream provides a straightforward means of analysis even at low mercury concentrations. Samples up to 200 µl can be introduced into the flowing stream without any change in the background level or loss of mercury. Furthermore, if a standard is introduced into a second loop on the same injector valve out of phase with the sample, then a sample/standard sequence can be automated and applied to on-line applications. Any baseline drift or calibration drift can easily be corrected in this approach.

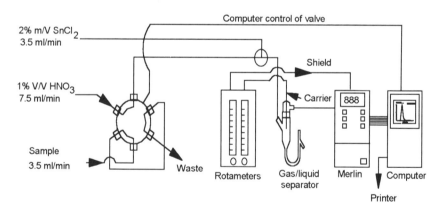

Fig. 5.7   Flow injection design described by Corns et al. [24].

For applications where large concentrations of mercury are expected, the fluorescence system has some limitations. Self-absorption occurs when levels above 1 ppm are experienced in normal operation of the mercury system. Corns et al. [24] have shown that the use of discrete sampling with injections of 75-200 µl will provide a linear calibration up to a maximum of 10 ppm. A further advantage accrues from the fact that samples with concentrations of up to 100 ppm can be tolerated in the manifold system without carry-over to the next sample. For analytical applications using large mercury concentrations this offers an attractive option, since in the standard continuous flow approach a level of 100 ppm will cause increased background, which may only be removed by complete dismantling of the glassware and flow manifold.

Another interesting area of application is where flow systems are used in conjunction with the flow-injection approach to select particular components from a sample, i.e. organic or inorganic mercury samples, or to preconcentrate particular elements of interest. McLeod and Wei Jian [25] have described a manifold system for determining inorganic and organic mercury by atomic fluorescence. The manifold is shown in Fig. 5.8. If a dual configuration valve is used, the sample can be injected in the first phase. The organic segment is retained on the column and the inorganic mercury provides a single peak corresponding to its concentration. When the valve is reset, a segment of hydrochloric acid is injected which removes the organic mercury from the column. Bromate/bromide reagent and chloride streams are used to oxidize methyl mercury and generate elemental mercury, respectively. In this manner it is easy to establish the total mercury present and to speciate the organic and inorganic forms.

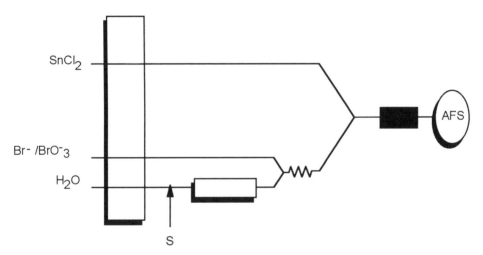

Fig. 5.8   Manifold design described by McLeod and Wei Jian [25].

Dean et al. [13] have described the characteristics of a flow-injection system linked to ICP-MS for trace metal analysis. They found that flow injection can be successfully used with ICP-MS and offers a number of advantages. Flow injection enables solutions with a high salt content, high viscosity and high acid strength to be successfully analysed for their elemental content without blockage or distortion of the sample interface. This is achieved by analyte dispersion occurring in the carrier tubing and decreased total loading into the plasma.

Flow-injection sample introduction has been successfully applied in the analysis of standard reference materials and in the measurement of accurate and precise isotope ratios, and, hence, isotope dilution analysis. The rapid sample throughput possible with FI should allow a four-fold increase in the sampling rate compared with conventional nebulization techniques.   Also, the amount of sample consumed per analytical measurement by FI is considerably less than continuous nebulization.   These considerations are of particular importance for the cost-effective operation of ICP-MS.

The reproducibilities of both the peak heights and peak areas are shown for replicate injections of a 100 ng ml^{-1} solution in Fig. 5.9. The results reveal that the injection volume has no significant effect on precision for the standard solution, providing the volume does not exceed 200 µl. Peak area generally offers greater precision (relative standard deviation 2%) at all injection volumes. As the injection volume is decreased, so the width at half-height decreases.

Other attempts at using such combinations include preconcentration of elements by using basic ion-exchange principles. Two methods of development are described here: firstly, the work by Bysouth *et al.* [26-28] and secondly that of Gunther Knapp at the University of Graz. These are now available commercially from Dionex, and from Knapp Logistics.

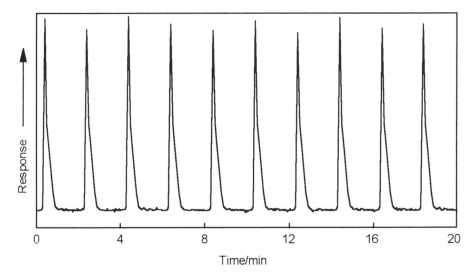

Fig. 5.9 An illustration of the reproducibility of 100 µl injections of a 100 ng ml^{-1} lead solution.

The measurement of very low levels of environmental pollutants is becoming increasingly important. The determination of lead, a cumulative toxin, is a good example. The current maximum allowable concentration of lead in British drinking water, before it enters the distribution network, is 50 ng ml^{-1} [29]. Although electrothermal atomization atomic-absorption spectrometry (AAS) can be used to measure this and lower concentrations, it is slow and requires considerable effort to ensure accurate results. Flames can provide simple and effective atom sources, but, if samples are aspirated directly, do not provide sufficient sensitivity. Thus, if a flame is to be used as the atom source, a preconcentration step is required.

Various methods of achieving preconcentration have been applied, including liquid-liquid extraction, precipitation, immobilization and electrodeposition. Most of these have been adapted to a flow-injection format for which retention on an immobilized reagent appears attractive. Solid, silica-based preconcentration media are easily handled [30-37], whereas resin-based materials tend to swell and may break up. Resins can be modified [38] by adsorption of a chelating agent to prevent this. Solids are easily incorporated into flow-injection manifolds as small columns [33,34,36,39,40]; 8-quinolinol immobilized on porous glass has often been used [33,34,36]. The flow-injection technique provides reproducible and easy sample handling, and the manifolds are easily interfaced with flame atomic absorption spectrometers.

Sec. 5.3] Flow-injection and preconcentration applications 149

The manifolds, which have been described previously, operate with injection of a large sample volume, either by timed flow-switching [33,36], or by using a large sample loop in an injection valve [34,39,40]. This second option allows only multiples of a discrete volume to be preconcentrated, unless the sample loop is changed. With timed injection, the preconcentration volume can, in theory, be infinitely varied. In some manifold designs, the column is placed just before the nebulizer of the atomic-absorption spectrometer so all the sample matrix and unadsorbed analyte will pass into the nebulizer during preconcentration. This could cause nebulizer or burner blockage, or an unstable baseline. However, these problems can be eliminated by diverting the stream away from the detector during preconcentration.

The manifolds described for preconcentration by Bysouth *et al.* [26] involve a column included within the sample loop of an injection valve. This enables timed sample loading onto the column without the matrix components passing to the spectrometer. Elution is achieved by switching the valve to place the column into the carrier stream which contains eluent. Four manifolds were used and these are shown in Fig. 5.10. Polytetrafluoroethylene (PTFE) tubing was used throughout the experiment. Manifolds 1-3 were used for preconcentration studies and were based on a commercially available autosampler which allowed the timing of external devices. Valves V_1 and V_2 were controlled by the autosampler.

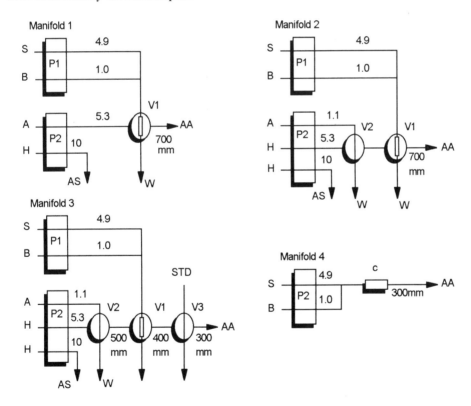

Fig. 5.10 Preconcentration manifolds: S, sample; B, buffer; A, acid; H, water; STD, standards; AS, autosampler probe wash-pot; W, waste; AA, spectrometer; C, column. Other symbols are explained within the text. Flow rates are in ml min^{-1}.

Valve V_1 had the column connected within the sample loop by using two 250 mm lengths of tubing, so that sample loading was done in the opposite flow direction to elution. The injection valve V_3 incorporated a 287 µl sample loop. Pump P_1 is switched off by the autosampler when the sample probe travels between the sample vial and probe washpot: this prevents air from entering the column. Pump P_2 runs continuously at a fixed speed.

The manifolds were used in the following manner. Sample and buffer were merged before being pumped to the column for 150 s, then valve V_1 was switched and the sample was eluted either by a continuous acid stream (manifold 1) or by an acid slug injected simultaneously via valve V_2 (manifolds 2 and 3). During elution, the sample probe resided in the washpot which contains water. This water was merged with buffer and washed the sample from the connecting tubing for 40 s. Valve V_3 (manifold 3) allowed the injection of standards that had concentrations above the normal detection limit of the instrument, whilst preconcentration was proceeding. In these manifolds, the elution flow rate was selected to give maximum signal for solutions injected without preconcentration. Manifold 4 was used to monitor the column effluent during preconcentration.

The manifolds gave accurate and precise preconcentration of lead, enabling the detection limits of flame AAS to be reduced by a factor dependent on preconcentration time. Placing the column within the sample loop enables a simple and effective manifold to be constructed without the sample matrix passing into the nebulizer. Manifold 1 is simple and effective, but consumes a considerable quantity of acid at the nebulization flow rate. When a second valve is included (manifold 2), the consumption of acid is reduced. When a third valve is included (manifold 3), other solutions can be injected during a preconcentration. Indeed, if a calibration is generated from the normal injection of standards, and the preconcentration factor is evaluated from one preconcentration standard, the system can quickly be calibrated. Each preconcentration of a standard takes a total of 190 s, compared with 7 s for a normal injection. The immobilized reagents used appear to be unselective for lead so that other species can compete for this reagent.

The use of masking agents in the determination of lead in tap-water by flame atomic absorption spectrometry was investigated by Bysouth *et al.* [28]. They showed that the selectivity of immobilized oxime for lead is improved by the use of masking agents during preconcentration, prior to determination by flame atomic-absorption spectrometry. Interference by iron, copper, aluminium and zinc are suppressed by including triethanolamine, thiourea, fluoride, acetylacetone or cyanide in the buffer as masking agents. Species such as iron or copper can completely prevent the preconcentration of lead. This was overcome by using a buffer consisting of 0.2 M boric acid, 2% triethanolamine, 2% thiourea and 2% acetylacetone, even when the interferent species is in a 200-fold excess over the lead. Recoveries from tap water samples, to which various amounts of lead were added, ranged from 94 to 108%. Results of analyses of tap water samples were in good agreement with those obtained by electrothermal atomic absorption spectrometry.

It is often the case that the approach to an automatic problem and the way it is solved is determined by the country of origin of its proposers. This is particularly true of the trace preconcentration system designed by Knapp *et al.* [41] and it represents the 'Rolls Royce' approach to analytical chemistry. It is not limited, it is totally flexible, and will, within reason, link onto any analytical instrument for measurement. That being said, it has found only limited application thus far, in that it does not provide any unique analytical methodology required by large groups of people. Perhaps this is because it has no real application area. It relies on the user to develop the necessary chemistry for the particular application. It is hoped that the commercial vendors of this instrument will be able to overcome this problem in the near future. In both on-line and off-line approaches it does seem to offer an attractive proposition. The solution might be to select a simple group of elements in a common matrix which must be analysed by several groups, perhaps trace elements in sea water samples.

To exploit the possibility of rapid multi-element determinations by modern sequential ICP instruments the pre-concentration procedure must be carried out off-line. Knapp et al. [41] have developed a microprocessor-controlled preconcentration instrument that performs various preconcentration regimes. In order to demonstrate the options available, trace elements have been concentrated in brines and in standard reference materials following digestion procedures. The determinations were carried out by sequential ICP-AES.

Figure 5.11 shows the flow diagram of the microprocessor-controlled pre-concentration equipment, which is configured here for off-line operation, and consists of a sample changer, three separate peristaltic pumps (P1, P2 and P3) for the sample solution, buffer and acid, three magnetic valves (V1, V2 and V3), the preconcentration column filled with chelating ion-exchange material (7 mm i.d., 10-30 mm height) and a fractionating unit for the acidic column eluate. The flow-rates for the sample solution, buffer and acid are adjusted to 5 ml/min.

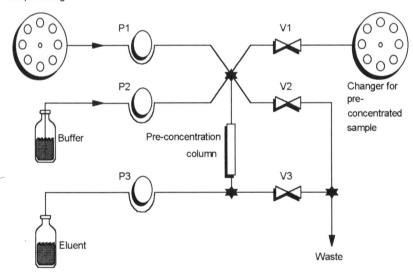

Fig. 5.11 Flow diagram of preconcentration system described by Knapp et al. [41].

Table 5.4 shows the preconcentration programme that was used in this investigation. Sample volumes of 10, 20 and 50 ml were used successfully and the enriched metals were in all three instances eluted with 5 ml of HNO_3 (2 M). The control unit is equipped with a CRT display to simplify the input of the desired preconcentration parameters. During the preconcentration, the current settings of valves and pumps are displayed.

Table 5.4   Preconcentration programme

Step	Status of pumps and valves*	Time/s	Comments
1	P1 + P2 - P3 - V1 - V2 - V3 +	12	Purge system with sample
2	P1 + P2 - P3 - V1 - V2 - V3 +	120 (cycle 1) 360 (cycle 2) 600 (cycle 3)	Preconcentration
3	P1 - P2 + P3 - V1 - V2 + V3 -	12	Purge system with buffer
4	P1 - P2 + P3 - V1 - V2 - V3 +	30	Column wash with buffer
5	P1 - P2 - P3 + V1 - V2 - V3 +	12	Purge system with acid
6	P1 - P2 - P3 + V1 + V2 - V3 -	60	Elution
7	P1 - P2 - P3 + V1 - V2 + V3 -	60	Column wash with acid
8	P1 - P2 + P3 - V1 - V2 + V3 -	12	Purge system with buffer
9	P1 - P2 + P3 - V1 - V2 - V3 +	120	Column condition
10	P1 - P2 - P3 - V1 - V2 - V3 - Start next cycle with step 1 (twice)	2	Actuate eluate changer
11	P1 - P2 - P3 - Start next sample with step 1	2	Actuate sample changer

* Pump status: - off, + on.   Valve status: - closed, + open.

5.3.1 Advantages of automatic preconcentration of elements with TraceCon

Preconcentration can enhance instrumental detection limits by two orders of magnitude. Interfering sample components (i.e. high salt contents) are readily removed. The commercially available TraceCon system provides a simple automated step for combining flame-AAS or ICP-OES analytical methods. Its robust design shows many good aspects of laboratory automation.

The TraceCon system achieves the portioning of the samples by a precision peristaltic pump. The sample volume is variable and not limited to the volume of a sample loop.

The ion-exchanger materials allow flow rates of 10 ml/min with minimum back pressure; even at these flow rates, 95-100% of the transition metals are retained on the preconcentration column. Throughput is maximized without detriment to the extraction efficiency.

Dispersion of elements during elution has been minimized by several measures, and especially good attention to detail. Preconcentration and elution take place in reverse direction. All connectors and tubes have a minimal internal volume. Fast exchange kinetics of the exchanger materials allow for higher flow rates during elution.

The TraceCon system is particularly suitable for automatic method development and the screen shown in Fig. 5.12 illustrates this point.

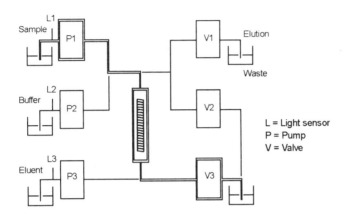

Fig. 5.12   TraceCon — a simple system for automatic method development.

For method development and display of the run status, a flow-chart of the TraceCon preconcentration system is used to visualize the current system state. During the execution of a method the flow chart displays the current activity of the TraceCon apparatus in the run mode. Run errors can also be easily identified, so that runs can be aborted if parts of the system fail.

The system has a wide range of applications and the approach offers many attractions, especially for ICP-AES and ICP-MS. However, to realize the full potential of the system, instrumental software must be provided to integrate these methods into the complete analytical system.

5.4   ON-LINE DILUTION SYSTEMS

The introduction of samples via nebulizers requires that they are either pneumatically or peristaltically pumped into the nebulizer for aerosol formation. This restricts the range of viscosities that can be easily handled by the nebulizer. For example highly saline or oil samples may well have to be diluted by an order of magnitude or greater. This dilution can be carried out either in a batch mode or continuously. Batch systems are quite complex in design but the rate of analysis is high. It is often the case that where dilution is required, in addition, a fast rate of analysis is also desirable. Some batch systems have been introduced commercially, notably to monitor wear metals in the oil industry.

An alternative approach has been developed by the Baird Corporation as a complete analytical system; this is also available as a stand-alone module from P.S. Analytical Ltd. The latter system is described here to illustrate the principles incorporated in the design. A block diagram of the system is shown in Fig. 5.13.

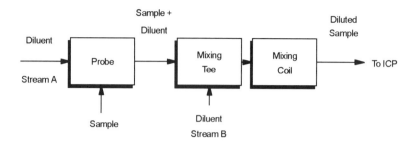

Fig. 5.13   Schematic block diagram of continuous-flow on-line dilution system.

It operates by the principle of differential pumping, using a series of peristaltic pumps, each of which operates under similar conditions. The pump tubing used on each pump is the same, and the variable-speed pump with which the initial dilution is provided will normally pump at full speed with the same flow rate as the other two pumps. The variable speed pump is set to operate at, for example, 80% of the fixed speed pumps. Diluent is pumped by this pump in to the dual concentric probe. This is subsequently withdrawn by the fixed speed pump. If the probe is in air, then segments of diluent interspersed with air are drawn through the probe by the fixed speed pump. When the probe is either in sample or wash, then liquid only is withdrawn from the probe with a dilution of approximately 1:5. The inner tube of the concentric probe is located so that dilution is provided almost as the sample is taken. In this way the viscosity range that can be analysed by the system is extended. Dilution takes place at the probe tip and therefore only diluted sample is withdrawn through the probe itself. A further dilution takes place at the NTM piece where the outputs of the two fixed speed pumps converge. The streams are further mixed in the coil and the stable diluted stream is passed over a sample pick-up well, at a flow rate configured by this, i.e. the flow rate of each fixed speed pump. This stream is then resampled at the flow rate required by the ICP or AA system to which it is linked. Figure 5.14 shows the pumping arrangement and the concentric probe in more detail.

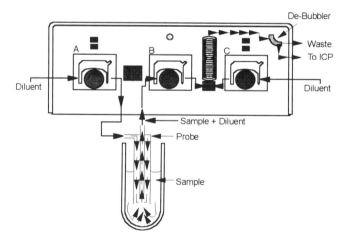

Fig. 5.14   Detail of concentric probe operation.

The analysis speed is determined (a) by the ICP or AA system flush-out characteristics; and (b) by the mixing and sample-transfer rates of the dilutor. The latter can readily be adjusted by increasing the flow-rate of the peristaltic pump tubes. The steady diluted stream can easily be adjusted in order that only a few seconds is required to flush out the previous sample and to set up a steady stable flow. Although peristaltic pumps are notorious for drift in flow-rate, the inclusion of an internal standard in the diluent can be a simple means of monitoring the flow change and of providing acceptable analytical results. Figure 5.15 shows the schematic arrangement of the concentric probe and the use of the probe wash pot. With viscous samples, the most likely cause of sample carry-over is from viscous oil collected on the probe exterior. To avoid this, the wash pot washes the probe exterior and interior, and flushes out the probe with diluent. This can lead to a considerable increase in sample rate. The probe and wash pot can be mounted on any commercially available autosampler.

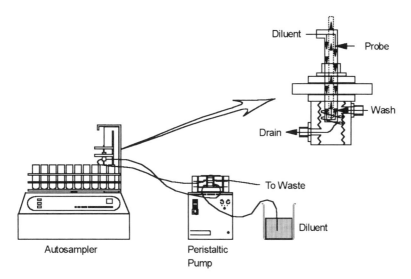

Fig. 5.15   Washpot and concentric probe schematic showing operation of unique washpot design.

A further advantage of this system is that it can easily be integrated with a commercial ICP-AES or ICP-MS instrument without software or electronic interface modifications. It will operate reliably for a wide range of viscosities and has applications not only in the oil industry but also for highly concentrated samples from other areas. It uses standard autoanalyser principles similar to the Technicon AutoAnalyzers.

5.5 SAMPLE INTRODUCTION INTO ICP-AES OR ICP-MS BY ELECTROTHERMAL VAPORIZATION

Although electrothermal vaporization has been widely accepted as an extension of atomic absorption, its use in inductively coupled plasma spectroscopy is fairly recent. In this technique the requirement for the vaporizer is somewhat different - the electrothermal vaporizer does not have to double as the atom cell. In fact, it is only needed to effect efficient and reproducible sample transfer from the rod, or a similar device, into the plasma.

Kleinmann and Svoboda [42] reported direct vaporization of samples into a low-power ICP source from a graphite disc support mounted directly within the body of the plasma torch. Nixon et al. [43] described the use of a tantalum filament electrothermal vaporization (ETV) apparatus as a sample introduction device for the inductively coupled plasma. They reported detection limits for 16 elements and pointed out that this ETV-ICP system gave an improvement in detection limits of between one and two orders of magnitude, in comparison with those obtained by using pneumatic nebulization of solutions into an inductively coupled argon plasma. This significant improvement in detection power has been attributed to the concentration of the analyte, already desolvated and vaporized by the tantalum filament, passing as a pulse of sample through the axial channel of the plasma and giving rise to a transient analyte atomic-emission signal.

Gunn et al. [44] described the application of a graphite-filament electrothermal vaporization apparatus as a sample introduction system for optical emission spectroscopy with an inductively coupled argon plasma source. Good detection levels were reported for the elements, and details of the interfacing requirements between the ICP and the graphite filament were explored.

The vaporization apparatus used a graphite rod (about 70 mm in length and 3 mm diameter) positioned between the terminals of the power supply in the shielded chamber; the terminals were cooled with tap water. The unit was contained within a cylindrical glass manifold (about 100 mm in diameter and 210 mm long). The total volume of the manifold was approximately 1 litre and the distance from the top of the manifold to the plasma was about 0.5 m. The enclosure was fitted with a conical top containing two ports. One had a ground glass ball joint, allowing the argon sweep gas carrying the vaporized sample to be transported to the injector tube of the ICP source. The other port was fitted with a polypropylene stopper and was positioned to allow the delivery by a micropipette of sample solution to the depression on the graphite rod. The filament was heated by a low-voltage, high-current power supply, which was fitted with a programmer allowing variation of the power and time during desolvation, ashing and vaporization procedures.

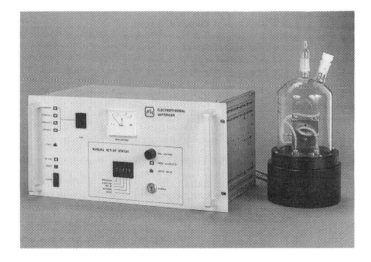

Fig. 5.16   Vaporization head design for linking to ICP/MS.

Figure 5.16 shows the vaporization head and the control electronics. The vaporizer uses commercial carbon rods readily available from Ringsdorf, Germany. The rod held in a gas-cooled mounting device is heated by using the control from the P.S. Analytical control module. The head is housed completely in a bell jar arrangement clamped on to a stand. This has two ports: one for injection of the sample with a long syringe, and the other as an outlet of argon gas to transport the sample. The argon flow over the electrothermal head can be directed into the torch continuously without any deleterious effect on the plasma. The preferred configuration is shown in Fig. 5.17. Here a valve is introduced so that the flushing argon from the electrothermal vaporizer head during the evaporation and drying stages can be vented to waste and only directed through the torch when actual analysis occurs. The bell jar arrangement introduces a large volume into the flow lines but this has no significant effect on the system (given adequate argon flow). A low flow will allow the sample, once volatile, to condense on the glass walls; reducing the volume of the bell jar will have a similar effect. A flow-rate of between 1 and 1.5 l min^{-1} of argon is sufficient to transport the vaporized samples into the plasma. Since there is a positive pressure of argon in the bell jar, and over the electrothermal rod, opening the cell to atmosphere will not affect the plasma's stability.

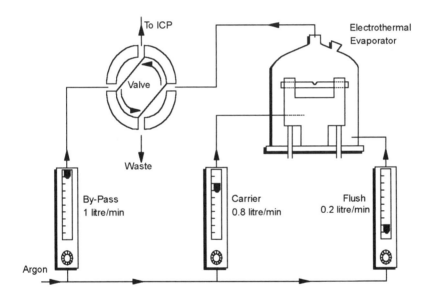

Fig. 5.17   Gas flow arrangement preferred for linking an electrothermal vaporizer to ICP/MS.

Rica *et al.* [46] evaluated the various parameters affecting the performance of the transport mechanism. Snook and Dean [47] have further extended this research. Both groups consider that the volume of the bell jar is important in the cell design.

The electronic controller is microprocessor-controlled and incorporates an RS 232 interface for linking to computers or extended devices. The controller allows four heating cycles to be set up with a time and voltage setting, to form a discrete program. A total of 10 programs can be entered into the local computer memory and run by simply starting the program by its program number.

With this instrument, the four heating cycles are identified by evaporation, pyrolysis 1 and pyrolysis 2, followed by vaporization. When a cycle is active, the LED opposite the stage is illuminated and the elapsed time remaining in this cycle is displayed in the two-digit LED register. The voltage directed to the rod in this cycle is indicated on the meter. On completion of the cycle time, the controller steps directly to the next cycle and proceeds throughout the program. A cycle within the program can be set to a time between 0 and 99 seconds.

The mode of operation of a graphite-rod electrothermal vaporization device for introduction of sample into an inductively coupled plasma source is different from that required in the use of this type of device in atomic-absorption spectrophotometry. The electrothermal device is required only to release a discrete pulse of analyte material from the sample transferred to it; this may be released either alone or with the components of the sample matrix and may also be released bound in molecular form, as finely divided particulate material or as free atoms. Atomization (and excitation for optical emission spectrometry) is then provided by passage of the analyte through the axial channel of the core of the ICP source. Provided, therefore, that no problems arise from premature loss of analyte during the desolvation and/or ashing stages of the temperature cycle employed with the graphite rod device, the requirements for close control of the final temperature of vaporization used to remove different analyte elements from the graphite rod is not as critical as in atomic-absorption spectrophotometry. It is necessary to provide a sufficiently high heating rate to the rod to ensure a rapid rate of removal of analyte from the surface and a discrete pulse of sample material above the surface of the rod for transport to the plasma source for excitation. For most of the elements studied, a satisfactory compromise vaporization temperature for the graphite rod was found to be $2400^\circ C$. However, in order to attain high sensitivity and precision in the technique employed, careful optimization of the parameters controlling sample transport to the ICP source was more important.

To optimize the applicability of the electrothermal vaporization technique, the most critical requirement is the design of the sample transport mechanism. The sample must be fully vaporized without any decomposition, after desolvation and matrix degradation, and transferred into the plasma. Condensation on the vessel walls or tubing must be avoided and the flow must be slow enough for elements to be atomized efficiently in the plasma itself. A commercial electrothermal vaporizer should provide flexibility and allow the necessary sample pretreatment to introduce a clean sample into the plasma. Several commercial systems are now available, primarily for the newer technique of inductively coupled plasma mass spectroscopy. These are often extremely expensive, so home built or cheaper systems may initially seem attractive. However, the cost of any software and hardware interfacing to couple to the existing instrument should not be underestimated.

Figure 5.18(a) shows a typical response cycle for an NBS sample of lead. The signal produced is obtained from 5 µl of solution injected onto the rod. The transient signal and the short time-scale from peak introduction to decay to zero level is clearly shown. The repeatability of the injection process, and the subsequent vaporization and measurement by ICP-MS, are shown in Fig. 5.18(b). This indicates repeat integrations, again using the NB 981 sample on the 208 Pb line.

Electrothermal vaporization has many applications relating to ICP analysis: the ability to vaporize solvents and to handle small solid or viscous samples are particularly important. In addition, the measurement of the isotopic ratios of lead in various ores provides a unique identification of the ore and its source. Now that commercial systems are available, the technique will undoubtedly become popular. These techniques seem to have found favour, particularly in Japan.

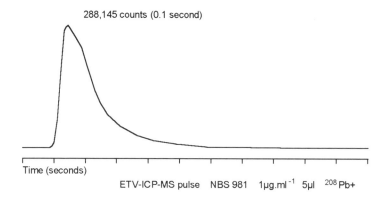

Fig. 5.18(a)   Signal obtained from 5 µl of NBS sample of lead.

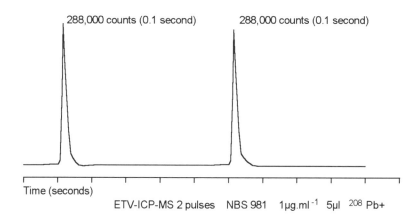

Fig 5.18(b)   Repeat injections of 5 µl NBS 981 - using 208 Pb line.

REFERENCES

[1] Backstrom, K., Gustavsson, A. and Hietala, P., *Spectrochimica Acta*, 1989, 44B, 1041.
[2] Gustavsson A., *Spectrochimica Acta*, 1987, 42B, 111.
[3] Gustavsson A. and Nygren, O., *Spectrochimica Acta*, 1987, 42B, 883.
[4] Gustavsson, A., *Spectrochimica Acta*, 1988, 43B, 917.
[5] Nygren, O., Nilsson, C.A. and Gustavsson, A., *Analyst*, 1988, 113, 591.
[6] Nygren, O., Nilsson, C.A. and Gustavsson, A., *Spectrochimica Acta*, 1989, 44B, 589.
[7] Gustavsson, A., *Trends in Analytical Chemistry*, 1989, 8 (9), 336.
[8] Montaser, A. and Golightly, D.W., *Inductively Coupled Plasmas in Analytical Atomic Spectroscopy*, New York, 1987.

References

[9] Stockwell, P.B., *Laboratory Practice*, 1990, 39 (6), 29.
[10] Goulden, P.D. and Brooksbank, P., *Analytical Chemistry*, 1974, 46, 1431.
[11] Dennis, A.L. and Porter, D.G., *Journal of Automatic Chemistry*, 1980, 2, 134.
[12] Godden, R.G. and Stockwell, P.B., *Journal of Analytical Atomic Spectrometry*, 1989, 4, 301.
[13] Dean, J.R., Ebdon, L.C., Crews, H.M. and Massey, R.C., *Journal of Analytical Atomic Spectrometry*, 1988, 3, 349.
[14] Branch, S., Corns, W.T., Ebdon, L.C., Hill, S. and O'Neill, P., *Journal of Analytical Atomic Spectrometry*, 1991, 6, 155.
[15] Cave, M.R. and Green, K.A., *Atomic Spectroscopy*, 1988, 9, 149.
[16] Hitchen, P., Hutton, R. and Tye, C., *Journal of Automatic Chemistry*, 1992, 14, 17.
[17] Thompson, K.C. and Reynolds, R.J., *Atomic Absorption, Fluorescence and Flame Emission - A Practical Approach*, Charles Griffen and Co., 1978.
[18] Thompson, K.C. and Godden, R.G., *Analyst*, 1975, 100, 544.
[19] Cossa, D., Personal communication, Ifremer, rue de L'Ile d'Yeu, BP 1105, 44311 Nantes, Cedex 03, France.
[20] Ebdon, L.C., Corns, W.T., Stockwell, P.B. and Stockwell, P.M., *Journal of Automatic Chemistry*, 1990, 11, 267.
[21] Stockwell, P.B., Rabl, P. and Paffrath, M., *Process Control and Quality*, 1991, 1, 293.
[22] Tyson, J.F., *Analyst*, 1985, 110, 419.
[23] Tyson, J.F., Adeeyinwo, C.E., Appleton, J.M.H., Bysouth, S.R., Idris, A.B. and Sarkissian, L.L., *Analyst*, 1985, 110, 487.
[24] Corns, W.T., Ebdon, L.C., Hill, S.J. and Stockwell, P.B., *Journal of Automatic Chemistry*, 1991, 13, 267.
[25] McLeod, C.W. and Wei Jian, *Spectroscopy World*, 1990, 2 (1), 32.
[26] Bysouth, S.R., Tyson, J.F. and Stockwell, P.B., *Analytica Chimica Acta*, 1988, 214, 329.
[27] Bysouth, S.R., Tyson J.F. and Stockwell, P.B., *Journal of Automatic Chemistry*, 1989, 11, 36.
[28] Bysouth, S.R., Tyson, J.F. and Stockwell, P.B., *Analyst*, 1990, 115, 571.
[29] Thompson, K.C., Personal communication, 1989.
[30] Hill, J.M., *Journal of Chromatography*, 1973, 76, 455.
[31] Leyden, D.E. and Luttrell, G.H., *Analytical Chemistry*, 1975, 47, 1612.
[32] Leyden, D.E., Luttrell, G.H., Sloan, A.E. and De Angelis, N.J., *Analytica Chimica Acta*, 1976, 84, 97.
[33] Malamas, F., Bengtsson, M. and Johansson, G., *Analytica Chimica Acta*, 1984, 160, 1.
[34] Marshall, M.A. and Mottola, H.A., *Analytical Chemistry*, 1985, 57, 729.
[35] Syty, A., Christensen, R.G. and Rains, T.C., *Atomic Spectroscopy*, 1986, 7, 89.
[36] Risinger, L., *Analytica Chimica Acta*, 1986, 7, 89.
[37] Allen, E.A., Boardman, M.C. and Plunkett, B.A., *Analytica Chimica Acta*, 1987, 196, 323.
[38] Brajter, K. and Dabek-Zlotorynska, E., *Fresenius Z. Analytical Chemistry*, 1987, 326, 763.
[39] Olsen, S., *Dansk Kemi*, 1983, 33, 68.
[40] Olsen, S., Pessenda, L.C.R., Ruzicka, J. and Hansen, E.H., *Analyst*, 1983, 108, 905.

[41]	Knapp, G., Muller, K., Strunz, M. and Wegscheider, W., *Journal of Analytical Atomic Spectrometry*, 1987, 2, 611.
[42]	Kleinmann, I. and Svoboda, V., *Analytical Chemistry*, 1974, 46, 210.
[43]	Nixon, D.E., Fassel, V.A. and Kniseky, R.N., *Analytical Chemistry*, 1974, 46, 210.
[44]	Gunn, A.M., Millard, D.L. and Kirkbright, G.F., *Analyst*, 1978, 163, 1066.
[45]	Long, S.E., Snook, R.D and Browner, R.F., *Spectrochimica Acta*, 1985, 40B, 553.
[46]	Rica, C.C., Kirkbright, G.F. and Snook, R.D., *Atomic Spectrocsopy*, 1981, 2 (6), 172.
[47]	Snook, R.D and Dean, J.R., *Journal of Analytical Atomic Spectrometry*, 1986, 1, 461.

6

Robots and computers

In the 1970s the future looked particularly rosy for developers of robotic systems. With hindsight it is clear that this was because something new really stimulates the users of automation; their enthusiasm has not been rewarded with the introduction of a vast number of robotic solutions to automatic problems. Indeed, the thinking now is away from the use of complicated robotic systems to more simple, uniformly usable systems. In this chapter a number of robotic systems will be reviewed and a path for future developments set out for the reader to ponder.

6.1 INTRODUCTION

The robot and robotics have been subject of countless publications. It is difficult, however, to obtain a clear picture of their acceptability: some see the robot as an indispensible technological tool for the 21st century; others attack it as a cause of unemployment. The development of robotics has been much less rapid than experts had expected early in the 1980s. What are the reasons for this slow-down: prohibitive costs, technological difficulties or organizational obstacles?

The introduction of the technological innovations provided by robotics has been the subject of many studies. Cabridain, for example [1] concluded that from the evidence of a study relating to a factory robotization:

> 'The people who conceive and design the implementation of new technology are far removed from the problems of everyday production. They complain that the new installations are not adequately adapted to the plant operation. However, the designers are not familiar enough with practical solutions. It seems as if every party involved in change in this case, the personnel in design and engineering, and the plant management each have a fragmented and partial understanding of the problems confronting the other. In practice, the decision-making process does not promote the exchange of information among all concerned and this in itself hinders the introduction of new technology'.

Whilst this study was not directly involved in analytical chemistry the lessons are clearly applicable.

In an attempt to understand the factors which lead to a successful and acceptable implementation of a robotic system, Molet et al. [2] have described the history of the implementation of a robotic installation in the steel industry in France. They analysed the reactions of the various participants. Most of the difficulties encountered were related to the technological modifications required in the plant to install the robot and its peripheral equipment, and the labour and organizational changes which affected those most directly involved. This experience provides some general lessons about the 'dos' and 'do nots' of robotization (see also Besson [3] and Guest [4]).

From inception to final implementation, robotic projects must involve everyone concerned. In the French project (undertaken at the Ecole National Supérieure des Mines de Paris in the early 1980s), the initiative to install the robot was taken by management and was not shared by other parties to the project. This separation of functions contributed significantly to the problems encountered (for example unexpected breakdowns, difficulties with peripheral equipment, staffing problems). Similar results have occurred in projects which were initiated and undertaken by the plant management, without the co-operation of other departments. It is not the competence of a particular function or service which is lacking, but, rather, their interaction. This is because:

> 'The robot can never be considered as an isolated machine in a plant: it is always an integral part of a production system. A project of robotization must emphasize the study of the robot and its environment, as an integrated whole, rather than relying on the consideration of the robot as a single machine. This notion, which integrates robot and peripheral equipment, must replace the simplistic idea that the insertion of a robot constitutes one more link, in the continuous technical evolution which has replaced man by machine, albeit this time by a very sophisticated one, of enhanced reliability and endurance. Unfortunately, it is this simplistic attitude that seems to pervade most enterprises which embark on robotization projects' [2].

The robot, by its very nature and attributes, is a machine which is most adaptable to a stable and deterministic environment. However, the real world of industry is far from being stable or deterministic. On the contrary, it is riddled with uncertainties. Some are short term and unexpected, like peripheral equipment stoppages, shortages of supply parts, dysfunctions in the robot memory. Some are longer-term uncertainties, such as those due to modifications in production methods needed to accommodate new product designs and replacement of knowledgeable and trained personnel. Training sessions and meetings can be held during a robotization project to try to remedy problems as they arise. These efforts represent only partial solutions to the whole process of adoption of new technology. Isolated actions from either the maintenance or the methods department cannot, by themselves, bring a project of robotization to life. The enterprise will only be in a position to take advantage of, and to master, the newly created 'flexibility' of incipient technologies, through concerted and pre-designed management policies that pool all resources, in the creation of a collective expertise.

Commitment to a robotic project by all involved (from the users to the top management) is as important in the analytical laboratory as it is in the steel industry. A robotization project needs a strong desire to succeed; this is very little different to any other kind of automation (see Chapters 1 and 2).

Robotics has been defined by Zenie [5] as an extension of programmable computers to do physical work, as well as processing data. Instrument systems using robotization and programmable computers are currently being used to improve productivity in scientific laboratories. Analysts need to identify samples, weigh, dilute, concentrate, extract, filter, evaporate, manipulate and analyse them.

6.2 ROBOT TYPES AND FUNCTIONS

The first commercial laboratory robot, the Zymate Laboratory Automation System (Fig. 6.1), was introduced by Zymark Corporation (Hopkinton, Massachusetts, USA) in 1982.

Subsequently, some light industrial robots have been adapted for laboratory use, and other systems have been introduced. Basic aspects of laboratory robots have been reviewed by Dessy [6,7], Kenig and Rudnic [8], Isenhour [9] and Lochmuller et al. [10]. More recently, Hawk and Kingston [11] have published a very comprehensive review with particular regard to trace analysis.

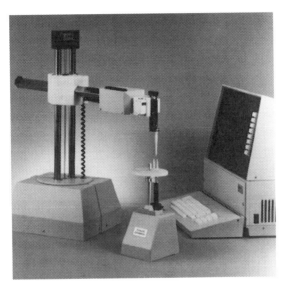

Fig. 6.1   The first Zymark laboratory robot. *Reproduced with permission of Zymark Corporation.*

Isenhour [9] proposed that robots be divided into three categories based on their evolutionary characteristics: first generation, second generation, and third generation [12].

First generation robots have fixed programs and are best suited for repetitive tasks; they do not have the ability to adapt to environmental changes. Most industrial robots are first generation.

Second generation robots have sensors, such as cameras, pressure-sensitive or light-sensitive pads, that provide information about their changing external conditions. Sensor information allows the robot to adapt to small changes in its surroundings.

Third generation robots possess features of artificial intelligence. This generation of robots is able to respond to environmental changes by modifying its actions to solve problems.

Many automated analytical instruments have been used to relieve laboratory technicians from routine work and thus increase their productivity. These instruments are well suited to hospital and factory laboratories, where the same analyses are performed every day. In more sophisticated laboratories, especially research laboratories, where the day-to-day analyses change, a more versatile instrument is needed. Robots in these laboratories will solve the problems arising from the non-versatility of automated instruments. The ability of a robot to do repetitious or dangerous work, with little or no external intervention, allows almost continuous generation of data and thus increases productivity, while decreasing the costs associated with having a human do the same work.

Bunce *et al.* [13], in a review of the application of robotics to clinical chemistry, have also attempted to classify the main types of robot available in a simple fashion. For example, robots may be static (i.e. floor, bench or ceiling mounted) or mobile on a tracked system.

The mechanical functions of a robot can be likened to those of the human body. The basic element is the trunk, to which is attached one or two arms which in some instruments have elbow joints. A variety of accessories can be attached to the end of the arm, such as a hand, fingers or special devices for pipetting etc. The range of these accessories varies between manufacturers, but they are not necessarily interchangeable between different robots. The accessories available should therefore be taken into account before choosing a robot.

Broadly, robots can be classified into five types:

Revolute: A robot consisting of rotational members arranged in a human hand, arm, shoulder and trunk configuration (for example the Perkin Elmer Masterlab Robot).

Cylindrical: A robot consisting of a central vertical sliding and rotating unit into which slides a horizontal beam (for example the Zymark Zymate).

SCARA: Selectively Compliant Articulated Robot Arm. This is a robot consisting of a central vertical sliding unit to which is attached two horizontally rotating links (for example the Universal Machine Intelligence RTX Robot).

Cartesian: A robot consisting of a vertical sliding unit which is mounted on an x-axis slideway carried on a y-axis horizontal sliding beam (an experimental laboratory robot of this type has been developed by the Laboratory of the Government Chemist in London).

Polar: A robot consisting of an extending horizontal beam, pivoted vertically at one end, mounted on a rotating unit (for example the GEC Robot Systems RAMP 2000).

Table 6.1 lists some of the equipment currently available, along with the robot type.

In contrast to a typical industrial robot, the laboratory robot must be flexible and user-programmable. Many laboratory robots incorporate tactile sensing and other verification methods. However, vision capabilities are virtually nonexistent at this time. Laboratory robots range in price from $25,000 - $100,000 with typical system prices averaging $40,000 - $50,000.

Automation needs vary from laboratory to laboratory. Quality control laboratories might emphasize user simplicity and high throughput, while research laboratories might stress flexibility.

Biological matrices usually require sample preparation prior to trace element analysis. These sample preparations, such as acid digestion or extractions, are labour intensive and can benefit significantly from laboratory automation to reduce both time and manual manipulations.

Table 6.1 Commercially available laboratory robots

Robot	Type
Biomek 1000	Cartesian
Josco Smartarm	Revolute
MasterLab	Revolute
Maxx-5	Revolute
Microassay	Cylindrical
Microlab 2000	Cartesian
Minimover-5	Revolute
Mitsubishi Move Master	Revolute
Puma	Revolute
Robotic Sample Processor 510	Cartesian
SCL-770	Cartesian
Zymate	Cylindrical

For details of robot manufacturers see the Appendix to this chapter.

6.3 ADVANTAGES AND JUSTIFICATION OF ROBOTICS IN THE LABORATORY

Table 6.2 summarizes the benefits that robotics brings to the laboratory.

Table 6.2 Benefits of robotics to the laboratory

The advantages of using robots in the laboratory include:

Improved precision.

Reliable documentation.

Higher productivity.

Safer laboratories.

Faster sample turnaround.

Staff become bored performing repetitive tasks, whereas a reliable laboratory robot will perform procedures uniformly, eliminating human error. Automation permits the frequent use of replicates, standards and controls to verify precision, so the analyst and customer can have complete confidence in the results. Taylor *et al.* [14] have suggested that the number of blind duplicates sent to control laboratories by operating staff can be as high as 70% of total work load. Reliability and traceability are important benefits of automation, for example, the sample preparation and analytical data acquired by the robotic system can be documented on the system printer or transmitted to a laboratory computer for data reduction or permanent filing.

A laboratory robot can operate unattended for 24 hours a day, releasing skilled technicians and scientists for more important and challenging work. Process workers can often be trained to bring the samples to the robot for analysis on a static mode operation. This provides valuable results with a fast timescale.

Staff have time to develop new skills and assume greater responsibility by delegating boring work to the laboratory robot. These new 'career opportunities' can reduce personnel turnover and retraining costs.

Laboratory procedures can require the use of potentially hazardous materials. Laboratory robots can minimize human exposure to hazards. Conversely, human contamination of biologically sensitive materials can also be minimized by laboratory robots.

Unattended operation of the laboratory robot extends the working day and permits faster sample turnaround. In a manufacturing/quality control environment, faster release of product for shipment can greatly reduce inventory costs.

In the past, laboratories have justified the initial investment in dedicated automation on the basis of the large number of identical, repetitive operations carried out. Fixed or dedicated automation is utilized for large quantities of standard procedures, such as those found in manufacturing environments or in clinical laboratories. Fixed automation follows a predetermined sequence of steps to perform a defined procedure; although efficient, it can only perform one repetitive procedure. Robotics, however, can provide flexible automation to meet the changing needs typical of quality control and research laboratories. Flexible automation is programmed by individual users to perform multiple procedures, and can be quickly reprogrammed to accommodate new or revised procedures. In these situations, a careful assessment of the software overhead must be made before a decision to purchase is made.

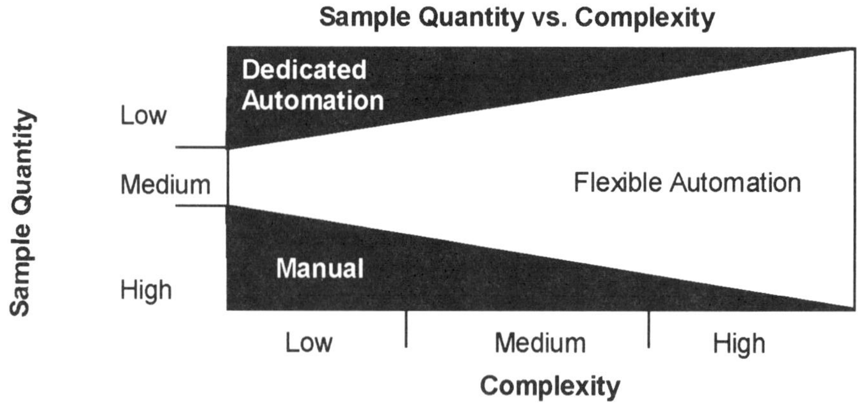

Fig. 6.2   Flexibility of automation offered by Hawk and Kingston [11].

The criteria for determining the feasibility of using flexible automation have been outlined by Hawk and Kingston [11] and are shown in Fig. 6.2. Sample quantity refers to the number of analyses using similar procedures. Complexity increases with the required precision, number of steps and critical timing requirements of the analytical procedure. Manual procedures usually provide the best approach for small sample numbers. Dedicated automation is more appropriate for high sample quantities of relatively low complexity. Flexible laboratory automation (robotics) fits into the region above manual techniques but below specialized, dedicated automation. Flexible automation methods can be used for dedicated applications. For example, the analytical services laboratory at a large chemical company introduced laboratory robotics for polymer testing. Each sample required multiple tests but no single determination required a sufficient amount of labour to justify dedicated automation. Laboratory robotics offered the flexibility to automate the determination of pH, viscosity and percentage solids on a single robotics system [15].The main difficulty in automating any procedure or operation is to identify the repetitive steps and to create a device that can accomplish them. The idea of automating entire laboratory procedures that appear to be unique to the compound and matrix being analysed may at first appear to be an impossible task. In practice, most laboratory procedures consist of sequences of common steps or building-blocks called laboratory unit operations (LUOs). Classes of LUOs have been defined - these are the building-blocks for all laboratory-scale operations and can be combined into multiple arrangements to meet specified requirements. Laboratory procedures can be systematically specified in terms of the required LUOs, their sequence in the procedure, and what happens at each LUO. Weighing, grinding, and liquid handling are examples of laboratory unit operations.

The most widely used laboratory robotic systems automate the individual LUOs with laboratory stations and use the robotic arm to transfer samples from one station to another according to user-programmed procedures. Samples are brought to each station, the controller sets all the necessary operating parameters (for example dispensing volumes, heating temperatures or mixing time). The controller waits for a signal from the laboratory station and then instructs the robot to move the sample to the next operation. The controller also sends control signals to analytical instruments and monitors status signals from these instruments in order to integrate the analysis and sample preparation steps.

Robotics, automated analytical instruments, and laboratory computers are now routinely used to automate each of the individual components of an analytical method. Complete automation of a method requires total integration of laboratory robotics, analytical instruments and computers. By integrating these components, samples can be prepared and analysed automatically. The resulting data can then be used to automatically control the remainder of the procedure. Hawk and Kingston [11] suggested that the goal of this type of integrated methods management system (Fig. 6.3) is to provide timely, quality decisions based on valid data. The ability to interact with a variety of standard laboratory instruments and computers greatly increases the flexibility of a laboratory robot.

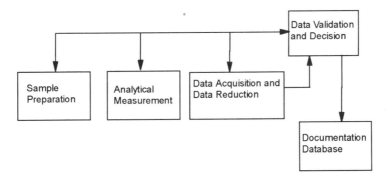

Fig. 6.3   Methods management schematic [11].

Current analytical instrumentation is capable of digital communication with other equipment such as computers and robots. The capability of remote operation is becoming a common feature of laboratory equipment, thus providing convenient interfaces with robotic equipment. This information and control capability is also leading to total laboratory information integration through the development of Laboratory Information Management Systems (LIMS). The trend toward co-ordination of laboratory information in an automated mode uses instrumental information capabilities and will create new opportunities for management through robotic operation of appropriate analytical equipment. However, there is still a considerable effort needed for the adequate definition of reliable, transferable interfaces.

In addition to improved productivity, quality of data is a key criterion on which many laboratory robotic applications are judged. Automated methods are generally validated by verifying that individual workstations are functioning properly, and then comparing the results of the totally automated procedure to those obtained manually.

A comparison of automated pipetting with manual pipetting, by Weltz [16], demonstrates a difference of approximately one order of magnitude; 0.4% RSD for automatic sample introduction versus 3.2% RSD for manual sample introduction. To achieve optimum results for trace elements in biological materials, the use of dedicated automated samplers is recommended. The liquid-handling capabilities of a pipetter/diluter workstation have been evaluated at Ortho Pharmaceutical (Haller *et al.* [17]). Four different volumes of each of three solvents were dispensed 20 times. Actual volumes dispensed were determined by mass. There was excellent agreement between the theoretical and actual results. Similar improvements have also been presented by Emmel and Nelson [18] for a robotic system to carry out classical wet digestion methods.

6.4 APPLICATIONS OF ROBOTICS

The applications of robotics in the laboratory have been numerous. The examples described here have been chosen from the recent literature to give a brief overview of what can be achieved using commercially available robotics, along with additional input from more conventional automation approaches.

Sample decomposition is not frequently automated because of difficulties caused by the long exposure times needed and the corrosive environments they produce. Hawk and Kingston [11] have described recent advances which make this preparative step more suitable for automation and particularly suited to robotic applications.

The evolution of closed-vessel microwave acid digestion for biological matrices reduces the time required for sample preparation to approximately 10 - 15 minutes, which makes it suitable for automation. The use of closed PFA Teflon containers reduces the amount of acid fumes which are corrosive to robotic equipment, and also reduces the possibility of contamination from the atmosphere. Kingston and Jassie [19] have shown that because the samples are digested at elevated temperatures and pressures, the time of completion of digestion of the sample is known. This is a requirement for robotically controlled equipment which cannot use an experienced chemist's visual examination to decide when a sample is ready for analysis. The benefits of closed-vessel microwave digestion have also been demonstrated for biological, botanical, and other materials by Matthes *et al.* [20], Nadkarni [21], White and Douthit [22] and Kingston and Jassie [19].

The first robotically controlled microwave decomposition system was developed at Kidd Creek Mines in Canada. This first application described by Labrecque [23] was developed for the decomposition of geological samples in order to analyse the matrix and trace elements.

Both microwave closed-vessel dissolution and laboratory robotics are relatively new to the analytical laboratory. However, it is this marriage of new methods which provides useful combinations of flexible laboratory automation to meet a variety of individualized needs. Because of the large number of biological samples which are prepared for analysis each day, it is reasonable to assume that this type of innovative automation will be of great benefit. It should be evaluated for its ability to improve the preparation technology for trace element analysis of biological materials.

Lester *et al.* [24] have described a robotic system for the analysis of arsenic and selenium in human urine samples which demonstrates how robotics has been used to integrate sample preparations and instrument analysis of a biological matrix for trace elements. The robot is used to control the ashing, digestion, sample injection and operation of a hydride system and atomic absorption instrument, including the instrument calibration. The system, which routinely analyses both As and Se at ppb levels, is estimated to require only one tenth of the human interaction used in the manual procedures that it replaces. The results remain comparable to manual operation.

6.4.1 Automated grinding of geological samples

Wilson and McGregor [25] have described an automated system for grinding geological samples. A Zymate II robotic system was used with specially designed or modified hardware to process partially pulverized geologic samples; the system processes samples for 24 hours a day with an average sample grinding time of 12.4 minutes. A sample grinding period is followed by a clean out sequence of air purges, vacuuming and grinding for surface cleaning with quartz sand. The operator has only to adjust the grinding plates on the Bico vertical grinder at the beginning of each preparation interval. Studies conducted using variable amounts of cleaning sand between grinding of samples indicated that the adjustment interval can be extended to 50 samples (25 g of cleaning sand) using an acceptance criteria of 80% of sample passing 80 mesh. The processed samples are suitable for direct geochemical analysis using a variety of standard chemical digestions. Cross-contamination studies using soil, chromate, and galena/sphalerite samples have revealed that the grinding system is capable of clean-out efficiencies exceeding 99% using as little as 25 g of cleaning sand.

The automated grinding procedure combined a Zymate II robotic system and a commercially available Bico model 6R vertical grinder. The robotic portion of the grinding system includes six sample racks providing a total capacity of 90 samples. A schematic diagram of the robotic system is shown in Fig. 6.4.

The sample racks, cleaning sand dispenser, waste cleaning-sand receptacle and Zymate II controller were positioned on a 5 x 5 ft counter-top to provide full access by the robotic arm. The sand dispensing unit was operated through the Zymate II controller using an air valve, which lowered the trough on the dispensing unit to dispense cleaning sand into the cleaning-sand container. A commercially available wet/dry vacuum cleaner was used throughout the study to remove residual dust from the system. The on/off cycle of the vacuum cleaner was controlled by the Zymate II system through a separate 10 amp power controller. The Bico grinder, which is situated inside a conventional fume hood, was turned off at the end of the grinding interval using a commercially available 24-hour dial switch located on the outside of the hood.

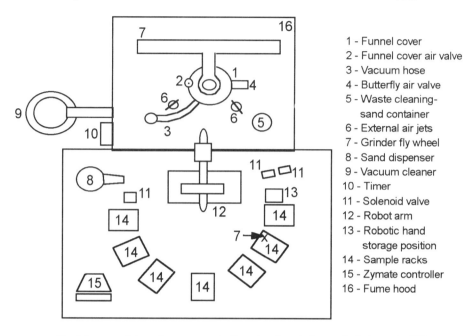

Fig. 6.4   Schematic arrangement of a system for grinding geological samples [25].

A model 6R Bico vertical grinder, equipped with a 220/240V three phase motor, was used in all studies. The grinding system was modified to incorporate a custom-designed hopper, hopper cover, sample-cup holder and cleaning jets, which were all co-ordinated by the Zymate II central controller.

The preparation sequence used in the robotic grinding of samples was similar to that used in the manual operation. The major problem was to set up adequate cleaning procedures before the next analytical sample.

The grinder is cleaned before the next sample by using a combination of high pressure (100 psi) air blasts, vacuuming and cleaning-sand. After the sample is ground, the grinder is purged of sample by a series of nine air jets located in the funnel cover and hopper. The jets deliver bursts of high pressure air expelling residual sample through the open butterfly valve. A second series of air purges, accompanied by a simultaneous vacuuming of the grinder through the sample cup holder, removes residual sample. In the next step, the robot arm grasps the cleaning-sand cup located in sample rack position 90. The container is filled with an quartz sand which is dispensed into the grinder using the same set of motions used for the original sample. The cleaning-sand scours the surface of the ceramic plates, removing any residual sample. After the cleaning-sand is dispensed, the robot, with assistance from the sample cup holder, positions the cleaning-sand cup under the grinder. The butterfly valve is opened and the ground cleaning-sand allowed to collect. After a four-minute grinding period, the cleaning sand cup containing the ground sand is removed from the grinder and emptied into a waste container. In the final cleaning step, any residual cleaning sand is removed from the grinder during the air purge/vacuuming sequence. The cleaning-sand sample cup is then refilled from the sand dispenser and returned to its starting location. Throughout the grinding sequence, the external parts of the grinder are subjected to a series of air blasts from carefully positioned air jets to minimize dust build-up on the outside, which may contribute to subsequent sample cross-contamination. The total grinding interval, including cleaning, is 12.4 minutes.

Wilson and McGregor [25] concluded that it is appropriate to use robotics for grinding materials for geological samples. Studies dealing with cross contamination revealed that even samples containing levels of Pb, Zn or Cr above 1% did not pose a problem, provided that 25 g of sand is used to clean the system prior to the next sample analysis.

6.4.2 Robotic automation of gravimetric analysis for environmental samples

Lindquist and Dias [26] have described a Zymark PyTechnology System which automates and integrates four methods: total dissolved solids (TDS) dried at 180°C, TDS residue on evaporation (TDS-ROE) dried at 103 - 105°C, total suspended solids (TSS) dried at 103 - 105°C, and total solids (TS) dried at 103 - 105°C. Approved Environmental Protection Agency (EPA) or equivalent methods are used. A TurboVap Evaporator has been incorporated into the design to speed up the sample evaporation.

The manual procedures for TDS, TDS-ROE, TSS and TS are described in 22 separate steps in Table 6.3. The automated procedures follow these same steps. The robotic system shown in Fig. 6.5 has a Zymate Laboratory Automation System at the heart of the system design.

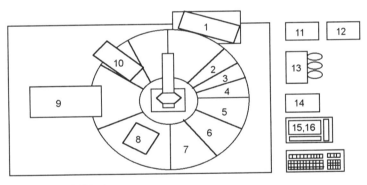

1 - Mettler AE 240 balance
2 - Filtration station
3 - Transfer station
4 - Hand station
5 to 7 - 12 position 250 ml fleaker racks
8 - 24 position Gooch custom oven
9 - Custom 3 tier dessicator - top
    24 position Gooch crucibles - middle
    12 position 100ml beakers - bottom

10 - Custom Turbovap - 9 position
11,12 - Power and event controller
13 - Master laboratory station
14 - Modumatic pump
15 - System V controller
16 - Computer

Fig. 6.5 Automatic system for gravimetric analysis [26]. *Reproduced with permission of Zymark Corporation.*

The major features of the automated procedure are:

1. Unattended operation and processing of 24 samples in 24 hours for TDS, TSS, TDS-ROE and TS.

2. Samples are stirred in 250 ml beakers during sample transfer using a magnetic stir bar.
3. Crucibles and beakers are cooled in a custom three-tiered desiccator.
4. An oven at 104°C, with flip-top lid, is used to dry the Gooch.
5. A TurboVap Workstation is used to dry beakers in an anodized aluminium heated block with a nitrogen purge at 104°C or 108°C.
6. The sample is transferred through a stainless steel cannula connected to a Teflon holding loop and Master Laboratory Station (MLS).
7. The filtration station can hold a Gooch crucible and/or beaker.
8. An electronic balance is used to weigh a Gooch crucible or beaker.
9. The filtration station cap has pin holes around the perimeter to allow residual sample drops to be rinsed out of Gooch crucibles with water.

Table 6.3 Manual methods for four solids procedures

Description	TDS	TDS-ROE	TSS	TS
1. Pre-wet filter	x	x	x	
2. Dry filter			104°C	
3. Dry dish	180°C	104°C		104°C
4. Cool filter in desiccator			x	
5. Cool dish in desiccator	x	x		x
6. Obtain constant tare weight	dish	dish	filter	dish
7. Transfer sample aliquot				x
8. Filter sample aliquot				x
9. Collect filtrate	x	x		x
10. Dry filter			104°C	
11. Dry dish	180°C	104°C		104°C
12. Cool filter in desiccator			x	
13. Cool dish in desiccator	x	x		x
14. Weigh filter			x	
15. Weigh dish	x	x		x
16. Re-dry filter			104°C	
17. Re-dry dish	180°C	104°C		104°C
18. Cool filter in desiccator			x	
19. Cool dish in desiccator	x	x		x
20. Weigh filter			x	
21. Weigh dish	x	x		x
22. If dry, then calculate results. If not, re-dry, cool and weigh.				

In order that the solids robotic system could be validated fully, each separate step or component of the procedure was isolated and the results verified gravimetrically. The validated results showed that the TS, TSS, and TDS-ROE robotic methods performed excellently in terms of precision, accuracy, and method detection limit (MDL). These results were found to be equivalent to, or better than, results obtained by manual methods. The TDS validation is in progress for the robotic method. Data for robotic and manual methods are set out in Table 6.4. for the TS, TSS and TDS-ROE methods.

Table 6.4   Validation of TS, TSS, TDS-ROE, TDS

Method	Robotic/ Manual	No. Meas.	Std. Dev.	Precision, rsd	Recovery	MDL mg/l
TS	Robotic	7	0.7	0.7%	99.9%	2
	Manual	7	2.2	0.6%	95.5%	7
TSS	Robotic	8	1.1	1.3%	102%	3
	Manual	8	0.7	0.9%	98.1%	2
TDS-ROE	Robotic	7	1.6	1.6%	102%	5
	Manual	7	1.5	0.3%	98.6%	5
TDS	Robotic	---------------- In-progress -------------				
	Manual	7	1.4	0.3%	100.3%	4

Using a Turbovap Evaporator for the evaporation of the TDS, TDS-ROE, and TS samples is extremely beneficial and these are summarized below.

1. It takes 30% less time to evaporate a sample to dryness at 104°C when using the gas flow on to the liquid in the beakers, compared with drying a sample using no gas flow.
2. The liquid is constantly mixed by the shearing action of the gas stream which reduces the likelihood of sample boil-over in the heated aluminium block.
3. Close temperature control can be maintained in order to achieve uniformity in sample processing.

The quality of the precision, accuracy, and MDL are comparable for the robotic and manual methods. A greater number of samples can be analysed in a faster turnaround time. Staff have also been freed from the mundane chores associated with these analyses for more demanding and rewarding tasks.

6.4.3 Robotic Karl Fischer titrator

Seiler and Martin [27] have described a robotic workcell for performing automated water determinations, which includes a Karl Fischer autotitrator and supporting hardware. The system is set up to process relatively non-viscous samples (common organic liquids) with water content falling in the nominal range 0.15-15%. The workcell is well suited to the industrial quality control laboratory, providing the following advantages:

1. Ease of use (minimal training/skill required for system use).
2. Adaptability (ability to handle routine or non-routine samples, changing priorities, replicates etc.).
3. High reliability (minimal downtime).
4. Capabilities to generate and use statistical data effectively.

The workcell is illustrated in Fig. 6.6. Most of the non-robotic components are located on a small table (about 60 x 150 cm) that stands adjacent to the robot table.

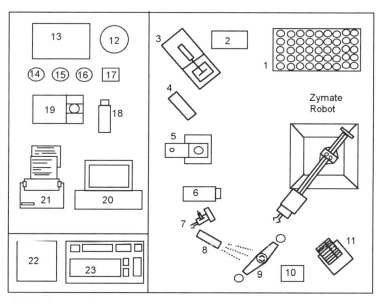

1 - Syringe rack
2 - Pump controller
3 - Mettler AE-200 balance
4 - Syringe inversion station
5 - KF titration cell
6 - Peristaltic pump
7 - Syringe/grip. comb. hand
8 - Barcode reader
9 - Capping station
10 - Grip. park
11 - Sample input rack
12 - Compressed gas cylinder
13 - Power/event controller
14 - T5, 15 - S1, 16 - S2, 17 - V1
18 - FMI metering pump
19 - Mettler DL-18
20 - Robot controller
21 - Printer
22 - Bar code printer
23 - Keyboard

Fig. 6.6   Schematic arrangement of work cell for Karl Fischer titrator [27].

These include an IBM PC/AT computer that functions in conjunction with a Zymate System V Controller located beneath the table top. Other devices located on this table include an autotitrator with solvent/titrant reservoirs, a pneumatic solvent switching valve, a metering pump, and a Zymark Power and Event Controller (PEC). A bar code printer provides labels for sample vials, and a printer is used for hard copy report generation. The devices mounted on the robot table (approximately 90 x 150 cm) are listed in Table 6.5.

The sample rack is unique in that it possesses 'pitch' both front to back and side to side. A single robot pick-up point is defined, and an optical (IR) sensor constantly monitors the pick-up point for the presence of a sample. Vials placed in the rack roll down to the pick-up point, under the influence of gravity. The rack allows implementation of a novel processing scheme.

All essential identification information related to the sample is either printed on the sample vial's bar code label or exists within a 'look up table' resident in controller memory. This approach avoids operator entry of this information through the computer keyboard and also facilitates implementation of a 'random order' processing scheme.

Many of the robotic manipulations used in this system (designed primarily for non-viscous samples) were adapted, in part, from those developed for a 'viscous liquids' KF system introduced by Zymark.

Table 6.5   **The Zymark (a) and other components (b) fixed to robotic table in the Seiler and Martin robotic workcell**

Zymark (a):
1. Robot arm
2. Standard gripper hand and park station
3. Capping station
4. Karl Fischer titration cell with pneumatic 'flapper' door and stirrer
5. Custom-built syringe/gripper combination hand and park station
6. Custom-built 'syringe inversion station'
7. Custom-built syringe rack

Other (b):
1. Analytical balance, equipped with a pneumatic door activation device from Zymark
2. Masterflex® peristaltic pump and controller
3. Laser bar-code scanner
4. Custom-built, gravity feed sample input rack, equipped with a narrow-beam infra-red (IR) source/detector pair.

A program called 'KF' controls all of the devices integrated into the system. It is written in Zymark's EasyLab control language, and, once started, does not stop until the operator aborts the run (or unless a fatal system error occurs).

The most important attribute of the robotic workcell is its ability to adapt to changing measurement situations. When sample processing priorities change, the operator can alter the run order without disturbing sample handling operations that might already be in progress. When a high precision result is needed, multiple determinations can be performed and averaged without significantly burdening laboratory staff. Another example of adaptability is the provision for automatic, or operator-specified, sample size selection based on anticipated results. The system can also accommodate viscous samples, as well as samples that yield deleterious side reactions with use of the usual methanol-based solvent. The system is very easy to use and requires no keyboard input after program start up.

6.4.4 Automated robotic extraction of proteins from plant tissues

Brumback [28] has described a custom robotic system designed to automate the extraction of proteins from plant samples. Leaf or callus material (5-25 mg) is presented to the robot in microcentrifuge tubes, the system performs buffer dispensing, grinding, centrifugation, and pipetting unit operations, and a cleared supernatant is delivered in a 96-well microassay plate format for subsequent analysis.

The system consists of two overhead X-Y-Z arms, an Allen-Bradley programmable logic controller, a microcomputer for user interface and control, and several custom peripheral devices for sample handling and grinding. The system is housed on a 120 x 150 cm table and contains a back-up power supply and a refrigeration unit to prevent sample degradation. The unit operates in a batch mode and is capable of processing more than 100 samples per hour.

Since the original procedures were long and tedious, Brumback [28] decided that the automation efforts must include procedural modifications to improve productivity and accommodate automation. The following design criteria for the system were designed:

1. Minimize and simplify procedural steps.
2. Batch operation with a minimum sample capacity of 100 per hour.
3. Minimize the amount of sample and reagents required, since all wastes involving recombinant organisms must be autoclaved.
4. Repeatability and uniformity should be equal to or better than the existing method.

A custom robotic system was developed to meet the design objectives. The peripherals accomplish four major LUOs: reagent delivery, sample grinding, centrifugation and sample pipetting. A schematic drawing of the system layout is shown in Fig. 6.7.

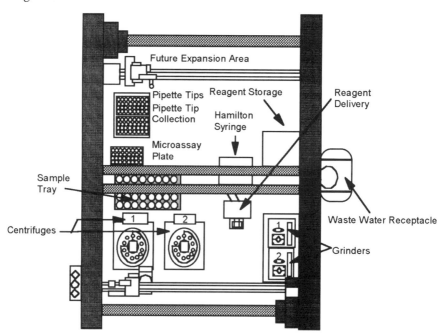

Fig. 6.7   Schematic diagram of robotic system for plant protein extraction.

Overall control of the system is accomplished by a COMPAQ 386s microcomputer running a control program written in Microsoft QuickBASIC. The two overhead robot arms are independently controlled by using individually programmable axis controllers in a daisy-chain configuration. Logic control and actuation of most peripherals is accomplished using a programmable logic controller (PLC). This contains three DC input modules, two AC and one DC output modules, and a BASIC language module to communicate with the microcomputer. PLC programming is accomplished by using ladder logic. Control of the programmable dispenser is provided by a direct RS-232-C communications link from the microcomputer. A schematic diagram of system control is shown in Fig. 6.8.

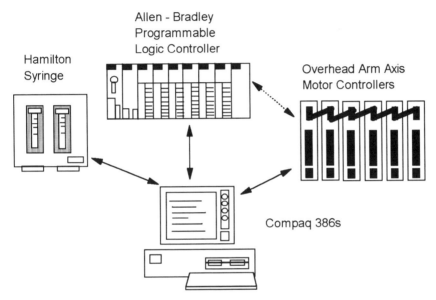

Fig. 6.8   Control of programmable dispenser via RS 232. *Reproduced with permission of Zymark Corporation.*

Subroutines within the overall QuickBASIC program monitor and control the communications ports. This configuration overcomes MS-DOS limitations and allows simultaneous communications across multiple communications ports. The multi-port access permits future expansion of the system, with other peripherals requiring communications with the control program.

Robot operation is controlled by user-created methods stored by the microcomputer. A method contains parameters that determine how samples are processed. Such parameters as centrifugation, grinding and wash times, reagent addition, pipette aspiration and delivery volumes, and save sample options can be specified in different methods depending on the application. The use of methods speeds up routine analyses by minimizing user inputs and interaction. For operation, the only inputs required are method name and the number of samples to be processed. The method development system can accommodate up to 100 different methods and contains options for creating, editing, deleting and copying methods.

The software system also contains a suite of password-protected maintenance programs. The unit is able to process 96 samples in 45 minutes or about 768 samples in an eight hour day.

6.5 ROBOTIC AUTOMATION OF TOTAL NITROGEN DETERMINATION BY CHEMILUMINESCENCE

Smith *et al.* [29] have described the evolution of the robotic system in organic liquids. Rapid, accurate, cost-effective analysis of liquid hydrocarbons for nitrogen and sulphur levels is a vital part of the smooth operation of an oil company. In 1984, a Zymate I arm and controller were purchased by Shell Development Co. for use in automating nitrogen determination. Some robotic systems are set up initially to perform their tasks and then operate unchanged, but the 'nitrogen robot' at Shell has taken eight years of continuous growth to reach its current design.

6.5.1 Background

The determination of total nitrogen by oxidative combustion with chemiluminescence detection is based on the following reactions:

(6.1) $\quad C_xH_yN + O_2 \rightarrow CO_2 + H_2O + NO + NO_2$

(6.2) $\quad 2\,NO + O_2 \leftrightarrow 2\,NO_2$

(6.3) $\quad NO + O_3 \rightarrow O_2 + NO_2^*$

(6.4) $\quad NO_2^* \rightarrow NO_2 + h\nu$ (detectable photon emitted)

As can be seen from reactions (6.3) and (6.4), only the NO formed during the combustion (equation 6.1) is quantified by the detector. The equilibrium shown in reaction (6.2) is affected by the temperature at which the combustion is dependent on the matrix of the sample being combusted. It is advantageous to match the matrix of the standards to the samples. Preparing standards to match a broad range of sample matrices is difficult and presupposes that the analyst already knows what the sample is, before analysing it. Alternatively, the matrix effects can be minimized by diluting the samples in the solvent used to prepare the standards. This not only takes care of matrix-matching samples and standards, but also can be used to extend the linear range of the analysis by diluting samples sufficiently to fall within the range of the standard curve.

The manual procedure for determination of total nitrogen by chemiluminescence involves preparing the standards from a stock solution, diluting the samples, injecting the standards into the combustion furnace for analysis, and calculation of the nitrogen content of the samples. The calculation is based on the signal from the diluted sample compared to the standard curve and the dilution factor for the sample.

During 1984, a Zymate I system was used to perform the dilution step of the total nitrogen determination. The system consisted of a Zymate I arm and system controller, a power and event controller, a balance for performing gravimetric dilutions, a vortex station for mixing the dilutions, a master lab station for dispensing the solvent, racks for sample vials, dilution vials and pipettes, a crimping station for attaching the dilution vial caps, a station for placing the diluted samples in an autosampler carousel and two detachable stands, one for pipeting and one for grasping and moving objects. The system had a 40-sample capacity and could be used for samples that were relatively non-volatile at room temperature. The system prepared a 1:5 dilution (one part sample with four parts solvent) of each sample using toluene as the solvent. The nitrogen robot handled 3000 samples per year.

By 1988 the nitrogen robot had undergone a number of changes in trying to address the limitations of the initial design. The system bore little resemblance to the initial system. In order to allow the system to handle a broader range of sample concentrations, additional dilution factors had been added. To maintain the precision of the dilution step, the size of the dilution vial had been increased from 12 x 31 mm to 20 x 40 mm. This allowed for up to a 1:99 dilution factor. To handle very viscous materials, such as tars and waxes, an aluminium heating block operating at $90^{\circ}C$ had been added to the system. The plastic pipette tips were replaced with heated glass pipette tips. Nitrogen standards were dispensed by the robot using the master lab station to calibrate the detector automatically. Shell cyclosol 63 replaced toluene as the diluent. The robot was linked directly to the pyroreactor and chemiluminescence detector so that if the diluted sample fell outside of the correct range, the sample could be re-prepped and run again, automatically, at the correct dilution factor. This required greater computer processing power than had been required for the initial system. This was obtained by the addition of an IBM XT computer that operated the pyroreactor, chemiluminescence detector and the Zymate II (upgraded from a Zymate I) arm and controller as peripheral devices.

The system in its current configuration is shown in Fig. 6.9. Since 1988, the system has been moved from a bench top to a custom vented enclosure to minimize exposure to organic fumes. The detectors, solvents and other necessary support devices are located below the robot work surface. Functionally, the robot still performs almost the same procedure as the initial system (dilution of samples), but there are a number of key differences.

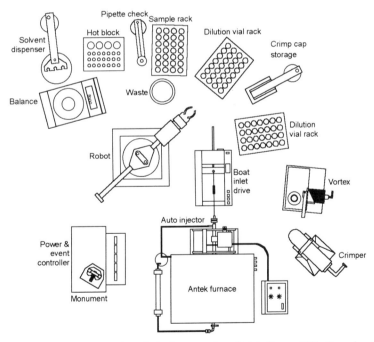

Fig. 6.9 System design of robotic automation of total nitrogen [29]. *Reproduced with permission of Zymark Corporation.*

Volatile samples are placed in the sample rack capped and are opened in a capping system only for long enough to transfer a sample aliquot to the tared dilution vial. An additional rack has been added to increase the 'on table' sample capacity of the system to 80 samples, but the software has been modified so that the racks loop back to the beginning when complete, giving the capability for an infinite number of samples per run. The empty dilution vials are taken from a custom vial dispenser with a capacity of 210 (on table) and infinite run capacity. After analysis, diluted samples are disposed of directly to a disposal container located below the table in the vented storage compartment. In addition to the heated glass pipette tips, there is a rack of room temperature glass pipette tips for use with non-heated and volatile samples. Error-checking software for the robot has been designed to diagnose and correct most common robot errors to maximize the ability of the system to keep running while unattended. The IBM software includes an automatic check for the quality assurance sample and electronic data archival and reporting. Two solvents are available, cyclosol 63 and toluene.

A second pyroreactor has been added to the system as back-up, to minimize the system down-time due to furnace heating element failure. The system has been expanded to also perform sulphur determination by oxidative combustion with UV fluorescence detection. The current sample load for the system is greater than 12 000 samples per year with a maximum capacity of the system, operating under optimum conditions, of greater than 40 000 samples per year.

The three systems are not distinct entities but, rather, represent a continuous evolution of hardware and software modifications over eight years. The system is vital to the operation of the laboratory, and it is not possible to have the nitrogen robot non-functional (for modification or repair) for more than one day.

6.6 A VALIDATED ROBOTIC SYSTEM FOR DISSOLUTION ASSAY OF AN IMMEDIATE-RELEASE/SUSTAINED-RELEASE TABLET

Dissolution testing, a routine requirement for tablets and capsules, is one of the most frequently performed tests in the pharmaceutical laboratory. It measures the dissolution of active drugs from solid dosage forms under standardized conditions. It is a tedious and labour-intensive task which has been successfully automated by the use of laboratory robotics. Guazzaroni et al. [30] have described the validation of a semi-automated procedure for USP dissolution testing of a tablet with an immediate-release active drug and a different sustained-release active drug. The robot was used to collect aliquots which were further assayed by HPLC. Acceptable linearities, reproducibilities and recoveries were established for the two active drug components. Also, the robotic and manual methods were compared.

A Zymate XP dissolution robot equipped with a balance and a six-spindle dissolution tester with remote interface was used. This equipment is shown in Fig. 6.10.

The aliquots collected by the robot were assayed by HPLC. The HPLC system consisted of an extended range LC pump, autosampler and an absorbance detector.

The dissolution method for the immediate-release/sustained-release tablet requires the following parameters: USP paddle method, 900 ml of water, 50 rpm paddle speed, $37°C$, and sampling points at 20 minutes, 40 minutes, 1, 2, 4, 6, 8 and 10 hours. Robotically, aliquots (8 ml) were removed, filtered through a 10 μm polyethylene filter and transferred to the storage rack. The volume (8 ml) was replaced with heated media. The samples were assayed by HPLC using the external standard method.

Two experiments were performed to validate the immediate-release component and the sustained-release component. To validate the sustained-release component, the

Sec. 6.7]   Automated extraction and analysis of chlorinated compounds   183

equivalent of one tablet weight of placebo blend and 100% label claim (LC) of the immediate-release drug substance were added to 900 ml of deionized water at 37°C in each of six dissolution vessels. The sustained-release drug substance was added to vessels 1 through 6 in increments from 10% to 120% LC. Aliquots were removed, filtered and stored for further HPLC analysis.

Fig. 6.10   Dissolution bench layout showing orientation of robot [30]. *Reproduced with permission of Zymark Corporation.*

To validate the immediate-release component, one table (containing 100% LC of the sustained-release drug substance) was added to 900 ml of deionized water at 37°C in each of six dissolution vessels. The immediate-release drug was added to vessels 1 through 6 in increments from 10% to 130% LC. Aliquots were removed, filtered and stored for future HPLC analysis.

Acceptable linearity, reproducibility and recovery was shown for the two active substances. Either peak-area or peak-height data may be used for quantification. Comparison of the manual and semi-automated dissolution tests have shown equivalency of the two methods.

6.7 AUTOMATED EXTRACTION AND ANALYSIS OF CHLORINATED COMPOUNDS BY LARGE VOLUME ON-COLUMN INJECTION GC/MSD

An automated system based on an XP Zymate laboratory robot, developed for the extraction and analysis of effluent water samples for the presence of 16 neutral chlorinated extractable compounds, has been described by Tamilarasan *et al.* [31]. The manual procedure involves the extraction of 400 ml of water sample with 40 ml of hexane, concentration of a 10 ml portion of the extract to 1 ml, and analysis of the concentrated extract by injection 5 μl into a GC equipped with dual analytical columns and dual electron capture (ECD) detectors. The manual preparation and analysis procedures were modified using the robotic system and a large volume on-column injection (LOCI) and gas chromatography/mass spectrometry (GC/MS) analysis technique. Thirty ml of the water sample is extracted with 5 ml of hexane and the extract is analysed by injecting 100 μl into a GC/MS. Injection of 100 μl of the extract into the GC/MS using LOCI, enabled the elimination tedious and time-consuming concentration step to be eliminated.

The automated robotic system performs the following steps:

1. Determine the sample weight and calculate the sample volume using a water density of 1.00 g/ml.
2. Uncap the vial and add 5 ml of hexane solvent (1:6 volume/volume ratio) which contain the internal standards.
3. Cap the vial and extract the analyte by mixing for 25 minutes using the linear shaker at a speed of 150 (available range 1 - 200).
4. Break the emulsion formed during shaking by vortexing in pulse mode for 10 minutes ('on' for 3 s and 'off' for 1 s).
5. Fill the LOCI 100 μl sample loop by sipping the extract through the sample loop using the sipper station.
6. Start the GC and initiate the sample injection and data collection.

The system is able to analyse a maximum of 40 samples per run, with an analysis time of 1 hour per sample. A sample is prepared while the preceding sample is being analysed.

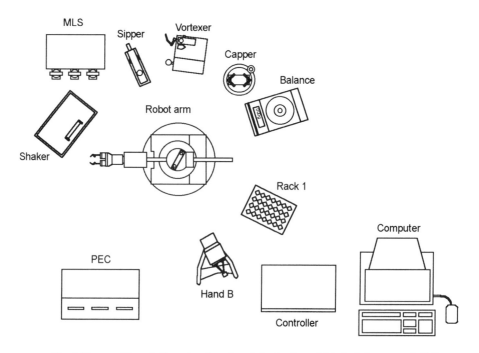

Fig. 6.11   The robotic system for automated extraction of chlorinated compounds [31].
Reproduced with permission of Zymark Corporation.

Sec. 6.7]   Automated extraction and analysis of chlorinated compounds   185

The system is based on an XP Zymate laboratory robot controlled with a 10 slot System V controller using software version XP V1.52. The system incorporates commercially available hardware, as well as custom hardware. A schematic diagram of the system is shown in Fig. 6.11. The robotic arm and the peripheral laboratory stations that the robotic arm interacts with to perform the application are positioned in a circular configuration. The GC/MS is located adjacent to the bench top, such that the injection valve is close to the sipper station. Peripheral items of hardware with which the robotic arm does not directly interact with are outside the working envelope.

The custom-made sipper station is used for transferring the hexane extract from the sample vial to the sample loop on the injection valve located on the GC/MS. The sipper station consists of a cannula and a washing shuttle. The cannula is a 14 cm long, 18-gauge needle mounted at the centre port of a two-port polypropylene circular disc connector. The side port of the connector is connected to the master laboratory station, through the sample loop on the injection valve by Teflon tubing. A schematic representation of the sample loop and its connections to the MLS and sipper station are shown in Fig. 6.12. The cannula is automatically moved up and down using a linear air actuator. The washing shuttle consists of a cylindrical polypropylene tower mounted on a platform. The washing shuttle is automatically moved in and out using a linear air actuator.

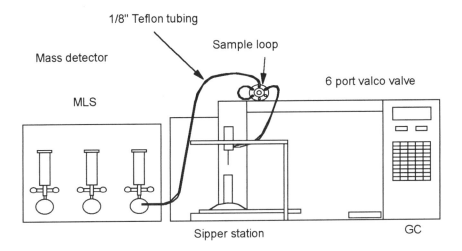

Fig. 6.12   The sample loop connected to MLS and sipper station [31].

The sample is extracted by shaking on the linear shaker. The linear shaker is equipped with a custom-made sample holder to hold the sample in a horizontal position.

The automated extraction procedure was optimized for maximum extraction efficiency. The extraction parameters investigated were shaking time, shaking speed and solvent/sample ratio. Three different shaking times (30 min, 60 min, and 120 min), two different shaking speeds (100 and 190) and two different solvent/sample ratios (0.25 and 0.1667) were examined. These parameters resulted in 12 possible sets of experimental conditions. Twelve spiked water samples were analysed in duplicate, in random order (selected using software), under these experimental conditions. The total volumes and the analyte concentrations were kept constant.

The automated robotic system was used to prepare and analyse more than 200 spiked water samples in order to evaluate the stability of the system. The system was very stable and no major problems were encountered during the stability studies.

The robotic automation of the procedure reduces the analysis time by a factor of >3 and the solvent consumption by 80 - 90%. The method detection limit and precision of the automated procedure are better than current manual procedures.

6.8 COLLECTION AND DETERMINATION OF DOSES DELIVERED THROUGH THE VALVES OF METERED AEROSOL PRODUCTS

The collection and analysis of doses from metered aerosol products has been found to be a tedious, labour-intensive process. Analyst-to-analyst variation in the actuating and collecting of samples from the canisters has been shown to be a significant variable in such determinations. Automation of the testing process conserves analyst time and eliminates all the variables associated with manual testing, thereby producing more consistent results.

Finley et al. [32] have developed a system to meet the various challenges offered by this automation, i.e. an actuator mechanism capable of both wasting and collecting metered doses. In addition, this actuator was required to shake the canisters and prevent waste actuations from entering the laboratory environment without the need for a dedicated fume hood.

A method to accommodate multiple waste/collection actuation protocols as well as differing canister sizes had to be designed. Also the option to weigh the canisters at various stages of the process was necessary.

Cost had to be contained within a tight budget and return on investment had to be realized. This was done by recycling existing equipment and using off-the-shelf parts wherever possible. Existing software was tailored to this application as appropriate. Data storage and retrieval to and from an existing, secured automated data system was required.

The metered dose aerosol is different from other dosage forms in that the reliable delivery of the specified dose to the patient is dependent upon the correct functioning of the entire delivery system throughout the claimed life of the canister. The metered dose valve, the only component of the system containing moving parts, must be tested to ensure accurate delivery for the canister life span. The canister is shaken immediately before each actuation, dispersing the active compound uniformly throughout the freon. Actuation is performed in the inverted position, with the valve filling for the next dose as the valve returns to its resting position. In the manual assay method, the canister is actuated in groups of 10 actuations with a 10 minute waiting period between actuation groups to allow the canister to return to thermal equilibrium following the cooling action of the release of the freon. With over 200 actuations per can for one formulation, this assay is tedious and time-consuming, requiring an average of about eight hours of analyst time to assay a group of six canisters. Automation of this method overcomes the wide variations that occur between individual analysts.

The system consists of a Zymate II Plus robot, a Compaq 386/16 PC, custom-built canister actuators, an analytical balance, a UV spectrophotometer and a pressurized tank and fill station, as well as racks for the canisters, beakers and standards. All activities of the system are co-ordinated and controlled by the PC under a scheduling prioritization algorithm. The movements of the robot are executed within the System V controller in response to parameters passed to it by the PC. The actuators are operated pneumatically, with positions of the pistons detected by magnetic sensors and driven by the PC. In order to meet the throughput demands, operations execute simultaneously whenever required. The system design is extremely flexible so that additional products and changes to methods can be implemented with minimal effects. A system configuration utility study has been included which allows the analyst to modify elements of the system. This prevents constant code changes to the software. Error checking facilities are also included.

Of particular interest to the operation of the system is the design of the actuator mechanism that shakes the canister, and performs both wasting and quantitative sample collection functions without active carry-over from one set of actuations to the next. Figure 6.13 describes the actuator and the components which collect the waste freon without the need for a separate fume hood.

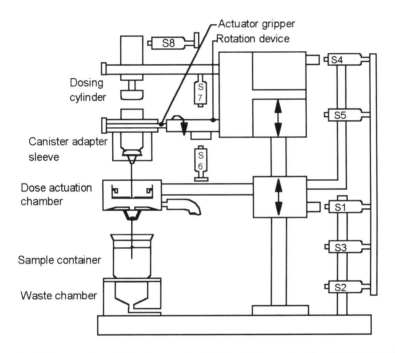

Fig. 6.13   InnovaSytems Inc.'s automated dose delivery system layout [32]. *Reproduced with permission of Zymark Corporation.*

The canister adapter sleeve must be placed on each canister before the system can transfer it from station to station. The canister adapter sleeve is made from Delrin and has an inside diameter sized to fit a particular aerosol container and valve. An O-ring or lip-type seal helps to centre and restrict relative motion of the canister within the sleeve. The outside shape and dimensions are constant for each canister size.

The actuation chamber is made from stainless steel. It houses the canister within its adapter sleeve during dosing. A smooth surface provides a face seal between canister valve stem tip and sample injection passage. The close tolerance between the external wall of the adapter and the internal wall of the actuation chamber induces a partial vacuum which creates a controlled pumping action causing sample methanol to flush the injection passage after each dose. A replaceable 16 gauge needle fits onto a standard luer tapered tip.

Internal passages and ports provide a means to rinse and air-dry the injection passage and the external wetted surface of the needle.

The dosing cylinder is a pneumatic cylinder which applies an adjustable downward force on the inverted canister to cause a dose to be expelled.

The actuator gripper is a holding device that accepts and positions the canister adapter sleeve during dosing and rotating.

The rotation mechanism rotates the gripper $90°$ to provide re-mixing of the medication within the canister.

Collecting a sample actuation from a canister is performed as a multi-step sequence. First, the canister is placed into an adapter sleeve. The sleeve, along with the canister, is then placed into the actuator gripper. The sleeve is then agitated by swivelling it to a horizontal position and back. Next, the canister adapter mechanism and dose actuation chamber are lowered. At this point the canister is positioned inside the dose actuation chamber and the dose actuation chamber rests on top of a sample container. The canister is then actuated by the dosing cylinder. Finally, the sleeve/canister assembly is raised from the dose actuation chamber and is ready to be re-actuated or unloaded.

Collecting a waste actuation from a canister is very similar to that of a sample actuation. The only difference is that the mechanism lowers into a waste chamber rather than a sample container. This waste chamber prevents waste actuations from entering the laboratory environment.

6.9 A SOLID REAGENT HANDLING ROBOT

The determination of the carbon and hydrogen content of liquid hydrocarbons requires a considerable amount of manpower. In the manual method, most of the analyst's time is spent sitting in front of the instrument, with an occasional break in preparing the water and carbon dioxide absorber flasks used in the procedure. An automated version of the method had been developed, utilizing several robot arms to manipulate the samples and the absorbers, and a computer to co-ordinate the activity of the arms and the combustion furnaces. The automation of the method has several immediate and significant effects. The precision of the method is improved by a factor of three, whilst the amount of manpower required to perform the analysis has dropped sharply. Turnaround time for the analysis decreased, since the robot can run for 24 hours per day, seven days a week. However, the sample load went up by a factor of three or more, so the number of absorbers used increased dramatically. The absorbers used for the manual method could be re-used 10 times, while the absorbers designed for use by the robot could only be used twice. So while the amount of manpower necessary to perform the analysis had decreased, the amount of time spent preparing the absorbers had increased. Absorber packing, which had been an almost desirable task with the manual method had become the scourge of the laboratory.

Smith *et al.* [33] described a robot system which was designed to address this sudden need for large numbers of absorbers. The system (shown in Fig. 6.14) is based on a Zymark Zymate II arm run with a System V controller, three power and event controllers and a balance interface with Mettler PE-160 balance).

The handling of air- and moisture-sensitive reagents and unusual reagent forms poses a number of challenges that have been addressed by this system. Means have been developed and tested for dispensing granular reagents, fibrous material and even mixing reagents at the time of use. The reagents are kept fresh in the storage bin and pre-set amounts are dispensed as needed.

To overcome the problems, all of the stations were custom designed and built in-house to handle the solid reagents and materials used to prepare the absorbers. The custom stations are: an absorber capping station, a glass wool dispensing station, three solid reagent dispensers and a station for positioning the glass wool and tamping down the reagents.

Fig. 6.14 System layout of solid reagent handling robot [33]. *Reproduced with permission of Zymark Corporation.*

The three solid reagent dispensers are particularly interesting and are described below.

There are two types of absorber necessary for the carbon/hydrogen determination; water absorbers and carbon dioxide absorbers. Figure 6.15 shows how each type of absorber is packed. For both absorber types, the absorber body is stainless steel tubing with the inside diameter ground and polished to allow a gas-tight O-ring seal with Teflon end caps. The end caps have a single hole in the centre to allow the combustion gases to pass through the absorber. A small plug of glass wool is placed in the ends of the tube to prevent the reagents from escaping through the hole in the end cap.

The water absorber is filled with a single reagent, magnesium perchlorate. The carbon dioxide absorber has four different layers of reagents in it, including one that is a mixture of two solids (4-10 mesh ascarite II and rock salt). The coarse ascarite II/rock salt layer needs to be mixed immediately before being placed in the absorber to avoid possible settling and separation of the mix. The exact quantity of each reagent present in the absorber is not critical, although too little could cause less than quantitative absorption of the combustion gases and too much would cause the absorber to overflow with reagents. Separate custom dispensers were designed and constructed to dispense each of the three reagents (magnesium perchlorate, 20-30 mesh ascarite II and 4-10 mesh ascarite II/rock salt) necessary to prepare the two absorber types. The dispensers use a shuttle to measure out the correct amount of reagent. When the shuttle is drawn into the fill position, the interior of the shuttle is filled with reagent from a nitrogen-purged reagent storage bin. When the shuttle is driven forward to the load position, the reagent drops out of the bottom of the shuttle chamber into the absorber positioned below.

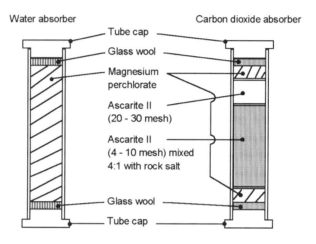

Fig. 6.15   Methods of packing for water and carbon dioxide absorbers [33]. *Reproduced with permission of Zymark Corporation.*

Any residual material left in the shuttle is blanketed by a slow flow of dry nitrogen gas to prevent fouling of the shuttle between uses. Pneumatic cylinders, driven from solenoid valves that are controlled by a power- and event-controller, provide the shuttle motion, and position is confirmed by Hall Effect detectors on the pneumatic cylinders and monitored by the power and event controller.

Both the water and carbon dioxide absorbers require magnesium perchlorate. For the carbon dioxide absorber there is a one-half inch layer at both ends, while the water absorber is completely filled with it. A station was designed for the magnesium perchlorate that could dispense two different volumes of material, one for the carbon dioxide absorber and a larger volume for the water absorber. The station uses two stacked shuttles to handle the two different volumes of reagent. For loading the one-half inch layer of material in the carbon dioxide absorber the upper shuttle of the station is filled and dispensed. To fill the water absorber with reagent, both upper and lower shuttles are filled from the reagent storage bin into the empty absorber tube.

The station designed to dispense and mix the 4-10 mesh ascarite II with rock salt uses a single shuttle, but the shuttle has two separate chambers in it. The shuttle is positioned under a divided reagent storage bin where one half contains 4-10 mesh ascarite II and the other half contains rock salt. When the shuttle is moved back to the fill position, the correct proportions of ascarite II and rock salt are loaded into the shuttle based on the size of the chamber allocated for each. When the shuttle is moved forward to the load position both chambers of the shuttle deposit their contents into a drop mixer. In the drop mixer, the reagents fall down a series of slides where each slide is in a different orientation to the one above it. The end result is that by the time the ascarite II and rock salt reach the bottom of the mixer (where the absorber is located) they have been adequately mixed. This arrangement has a definite advantage over dispensing a pre-mixed ascarite II/rock salt reagent as that would have a tendency to settle and separate upon standing in the storage bin. Correct mixing is critical to the functioning of the absorber.

The last and simplest of the three reagent dispensers is for the 20-30 mesh ascarite II. Only one volume of the material is used in preparing the absorbers so just a single shuttle with a single chamber is necessary for the station. As with the other two stations, the shuttle is filled with reagent from a dry nitrogen-purged storage bin and dispensed into the absorber tube positioned below the station by moving the shuttle forward.

An end cap dispenser, a glass wool dispenser and a plunger station were also designed and built.

6.10   CONCLUSIONS

The applications described here illustrate the wide range of uses for robotic systems. This chapter is not intended to be exhaustive; there are many other examples of successful applications, some of which are referenced below. For instance, Brodach *et al.* [34] have described the use of a single robot to automate the production of several positron-emitting radiopharmaceuticals; and Thompson *et al.* [35] have reported on a robotic sampler in operation in a radiochemical laboratory. Both of these applications have safety implications. Clinical applications are also important, and Castellani *et al.* [36] have described the use of robotic sample preparation for the immunochemical determination of cardiac isoenzymes. Lochmuller *et al.* [37], on the other hand, have used a robotic system to study reaction kinetics of esterification.

REFERENCES

[1]   Cabridain, M.O., *Automation and Organization of Work*, Annales des Mines, Gere et Comprendre, No 1, 4th Quarter 1985.

[2]   Molet, H., Sautory, J.C. and van Gigch J.P., *International Journal of Production Research*, 1989, 27, 529.

[3]   Besson, P., *Atelier de Demain (The Planet of Tomorrow)*, 1983, Paris (PUF).

[4]   Guest, R., *Robotics: The Human Dimension*, Pergamon, New York, USA, 1984.

[5]   Zenie, F., *Advances in Laboratory Automation Robotics*, Edited by Strimatics, J.R. and Hawk, G.L., Zymark Corporation, Hopkinton, MA, USA, 1984, 1.

[6]   Dessy, R., *Analytical Chemistry*, 1983, 55, 1100A.

[7]   Dessy, R., *Analytical Chemistry*, 1983, 55, 1232A.

[8]   Kenig, J.L. and Rudnic, E.M., *Pharmaceutical Technology*, 1986, 1, 28.

[9]   Isenhour, T.L., *Journal of Chemical Information and Computing*, 1985, 25, 292.

[10] Lochmuller, C.H., Custiman, M.R. and Lung, K.R., *Journal of Chromatographic Science*, 1985, 23, 429.

References

[11] Hawk, G.L. and Kingston, H.M., Laboratory robotics and trace analysis in *Quantitative Trace Analysis of Biological Materials*, Edited by McKenzie, H.A. and Smythe, L.E., Elsevier, Amsterdam, 1988.

[12] Vukobratovic, N., Stokic, D., *Scientific Fundamentals of Robotics 2, Control of Manipulation Robots*, Springer-Verlag, Berlin/New York, 1982.

[13] Bunce, R.A., Broughton, P.M.G., Browning, D.M., Gibbons, J.E.C. and Kricka, L.J., *Journal of Automatic Chemistry*, 1989, 11, 64.

[14] Taylor, G.L., Smith, T.R. and Kamla, G., *Journal of Automatic Chemistry*, 1991, 13, 3.

[15] Mango, P.A., *Advances in Laboratory Automation Robotics*, Edited by Strimatics, J.R. and Hawk, G.L., Zymark Corporation, Hopkinton, MA, USA, 1984, 17.

[16] Weltz, B., *Atomic Spectroscopy*, VCH Publishers, Deerfield, FL, USA, 1985.

[17] Haller, W., Halloran, J., Habarta, J. and Mason, W., in *Advances in Laboratory Automation Robotics*, Edited by Strimatics, J.R. and Hawk, G.L., Zymark Corporation, Hopkinton, MA, USA, 1986.

[18] Emmel, H.W. and Nelson, L.D. in *Advances in Laboratory Automation Robotics*, Edited by Strimatics, J.R. and Hawk, G.L., Zymark Corporation, Hopkinton, MA, USA, 1986.

[19] Kingston, H.M. and Jassie, L.J., *Analytical Chemistry*, 1986, 58, 2534.

[20] Matthes, S.A., Farrell, R.F. and Mackie, *Technical Progress Report No. 120*, US Bureau of Mines, Washington, USA, 1983.

[21] Nadkarni, R.A., *Analytical Chemistry*, 1984, 56, 2233.

[22] White, R.T. Jnr and Douthit, G.E., *Journal of the Association of Official Analytical Chemistry*, 1985, 68, 766.

[23] Labrecque, J., in *Microwave Sample Preparation - Theory and Practice*, Edited by Kingston, H.M. and Jassie, L.B., American Chemical Society, Washington D.C., USA, 1988.

[24] Lester, L., Lincoln, T. and Donoian, H., in *Advances in Laboratory Automation Robotics*, Edited by Strimatics, J.R. and Hawk, G.L., Zymark Corporation, Hopkinton, MA, USA, 1985.

[25] Wilson, S.A. and McGregor, R., *Analytical Letters*, 1989, 22, 647.

[26] Lindquist, L.W. and Dias, F.X., in *Advances in Laboratory Automation Robotics*, Edited by Strimatics, J.R. and Little, J.N., Zymark Corporation, Hopkinton, MA, USA, 1991.

[27] Seiler, B.D. and Martin, L.D. in *Advances in Laboratory Automation Robotics*, Edited by Strimatics, J.R. and Little, J.N., Zymark Corporation, Hopkinton, MA, USA, 1991.

[28] Brumback, T.B. Jnr., in *Advances in Laboratory Automation Robotics*, Edited by Strimatics, J.R. and Little, J.N., Zymark Corporation, Hopkinton, MA, USA, 1991.

[29] Smith,T.R. Kamla, G.J. and Kelly, M.P., *ISLAR 1992 Proceedings*, Zymark Corporation, Hopkinton, MA, USA, 1992, 670.

[30] Guazzaroni, M.E., Chen, Chone, N. and Bird, B.A., *ISLAR 1992 Proceedings*, Zymark Corporation, Hopkinton, MA, USA, 1992, 246.

[31] Tamilarasan, R., Morabito, P.L., Butt, A. and Hazelwood, P.D., *ISLAR 1993 Proceedings*, Zymark Corporation, Hopkinton, MA, USA, 1994, 396.

[32] Finley, T.A. Landes, N.A., Fuchs, R.M. and Zepka, A.J., *ISLAR 1993 Proceedings*, Zymark Corporation, Hopkinton, MA, USA, 1994, 163.

[33] Smith, T.R., Streetman, G.W. and Lopez, J.A., *ISLAR 1993 Proceedings*, Zymark Corporation, Hopkinton, MA, USA, 1994, 454.

[34] Brodach, J.W., Kilbourn, M.R. and Welch, M.J., *Applied Radiation Isotope, (Journal Radiation Applied Instrumentation Part A)*, 1988, 39, 689.
[35] Thompson, C.M., Sebesta, A. and Ehmann, W.D., *Journal of Radioanalytical and Nuclear Chemistry*, 1988, 124, 449.
[36] Castellani, W.J., Van Lente, F. and Chou, D., *Clinical Chemistry*, 1986, 32, 1672.
[37] Lochmuller, C.H., Lloyd, T.L., Lung, K.R., Gross, P.M. and Kaljurand, M., *Analytical Letters*, 1987, 20, 1237.

APPENDIX: ROBOTIC COMPANIES WITH PRODUCTS LISTED IN TABLE 6.1

Biomek 1000	Beckman Instruments 1050 Page Mill Road Palo Alto, California 94304, USA
Josco Smartarm	Johnson Scal Corporation 235 Fairfield Avenue West Caldwell, New Jersey 07006, USA
MasterLab	Perkin Elmer Corporation 761 Main Avenue Norwalk, Connecticut 06859, USA
Maxx-5	Fisher Scientific 711 Forbes Avenue Pittsburgh, Pennsylvania 15219, USA
Microassay	H.P. Genenchem 460 Pt. San Bruno Boulevard San Francisco, California 94080, USA
Microlab 2000	Hamilton Corporation P.O. Box 10030 Reno, Nevada 89520, USA
Minimover-5	Microbot Incorporated 453-H Ravendale Drive Mountain View, California 94043, USA
Mitsubishi Move Master	Hudson Robotics Incorporated 120 Morris Avenue Springfield, New Jersey 07081, USA
Puma	Unimation Incorporated Shelter Rock Lane Danbury, Connecticut 06810, USA
Robotic Sample Processor 510	TECAN US P.O. Box 2485 Chapel Hill, North Carolina 27515, USA
SCL-770	Seiko Instruments & Electronics Limited 6-31-1 Kameido, Koto-Ku Tokyo, Japan
Zymate	Zymark Corporation Zymark Center Hopkinton, Massachusetts 01748, USA

7

Examples of automatic systems

In this chapter a number of applications are described and the lessons that can be learnt from them about solving particular problems are highlighted. The examples do not necessarily present a complete solution to a problem, nor are the solutions described here the only means of solving a particular problem. However, each of the systems has been in routine use for a considerable time, and, although developed for specific applications, each has a wider degree of applicability.

7.1 AUTOMATIC SPECIFIC GRAVITY AND REFRACTIVE INDEX MEASUREMENT

In the fragrances and flavours industry, product quality is generally characterized by measuring density (specific gravity), refractive index and also occasionally, the colour index. Fragrances are characterized by a wide variation in viscosity and a tendency to have a repellant smell when concentrated. In situations where a variety of fragrances are used, the normal laboratory procedures can be time-consuming and the precision required relies upon considerable analytical dexterity and attention to detail. Until recently, there was no completely automated commercial system available. Figure 7.1 shows a system to measure both these parameters.

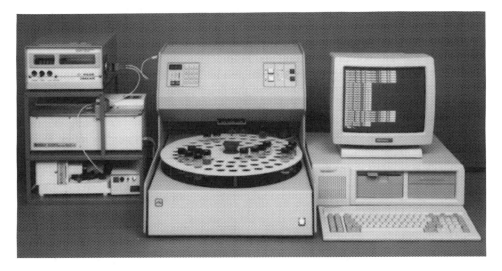

Fig. 7.1 Automated system to measure density and refractive index of fragrances and flavours.

7.1.1 Automated instrument design concepts

The samples are held in sealed vials so that the smells are retained. The autosampler allows, through an RS232 interface, individual selection of the container to be sampled. The control interface, itself a single-board computer, is also used to control the analytical sequence and to trigger the measurements. The vials contain samples of varying viscosity, and the process of taking the sample is both difficult and time-consuming. The actual sampling pump is located down-stream of the density-measuring instrument and the refractive-index device. In this way, the samples are not contaminated by the pump construction. A piston pump is chosen because this has a limited variation of flow rate due to viscosity change. A peristaltic pump would not be able to cope with the viscosity change in this application, and, if it did, would introduce variable time delays. Attention to detail is also required around the sampling probe and a specially-constructed device is used. This punctures the vial septum and allows air to replace the sample as it is withdrawn. Viscous sample held on the outside of the probe is then also stripped off, firstly by contact with the septum, and, secondly, with the special washpot.

The nature of the samples also makes it difficult to flush them from the system prior to analysis of the next sample. A combination of industrial methylated spirits (IMS) and air provides an acceptable solution. The flow rate of the sampling line is greater than the flow input of the IMS to the washpot, so alternate plugs of IMS and air are drawn down the Teflon tubing connecting the instruments in line. The physical plumbing and the orientation of the devices is also important. Too narrow a bore on the interconnecting tubing will cause additional back pressure and cause the pumping to fail. Too wide a tubing will result in unnecessarily long flush-out times and delay the analysis. The orientation of the instruments shown in Fig. 7.1 is the best compromise. This necessitates the construction of a simple instrument stand so that the distance between the input and output from the density measuring device and the refractive index instrument are minimized.

The sample size required depends on the viscosity of the sample, the cell size of the density measurement device and the refractive index instrument. The density measurement requires a sample size of about 7 - 10 ml and the measurement must be made after a stabilization period with no flow. In contrast, the refractive index measurement is made on a flowing stream.

In the system described here, the user had a product base of some 6000+ samples and a complete database of records of past history. The system specification called for the company's central control computer to be able to download a selection of samples for analysis, along with the permitted range of the density and refractive index data. The analyst would then load the autosampler in the correct pattern and allow an IBM PC to control the sequence of events, collect the data and display it for editing prior to transmitting it back to the central computer for archiving. The system therefore had to be specially designed with interfaces to the control computer and to each of the instruments it controlled and from which it collected data. The assistance of Allied Data Ltd, Hetton Lyons Industrial Estate, Hetton, Houghton-le-Spring, Tyne & Wear, UK, in this was greatly appreciated, since the three interfaces, although they were basically RS232 in nature, accepted completely different protocols.

The autosampler required a RS232 communication at 9600 baud using ASCII character strings in a question-and-answer manner. The density device (Anton Paar) communicated at 2400 baud and a complete data buffer of information was transmitted. The computer program was designed to select and validate data for each measurement.

Sec. 7.1]   Automatic specific gravity and refractive index measurement   197

Finally, the refractive index instrument had to be strobed to provide data using an RS232 interface at 1200 baud.

Once a series of samples is placed on the carousel, the analysis proceeds automatically with a cycle of approximately 3 minutes (including complete wash-out of the previous samples). When a representative sample fills the two detector cells, the flow is stopped and the measurements are collected from the instruments. When stable readings are obtained, they are compared with the data on file for the sample type and then presented to the analyst for acceptance. Measurements for colour are also possible with suitable changes in the design. Figure 7.2 shows the communication lines required and the protocols necessary to consolidate and operate the system. Such a system has been in operation on a routine basis for several years in a major fragrance and flavour company in the UK.

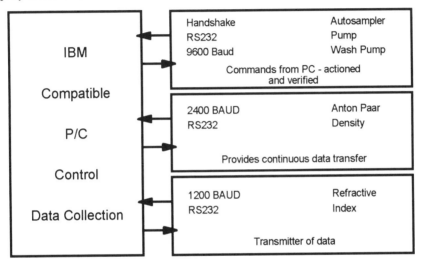

Fig. 7.2   Communication between three or four manufacturers' instruments and IBM compatible computers.

Table 7.1 sets out the problems involved in analysing fragrances and flavours and the advantages of the system that was developed.

Table 7.1   Problems encountered in the analysis of fragrances and flavours and the advantages obtained with a well designed automated instrument

Problems	Advantages
Highly viscous samples	Hands-off analysis
Toxic	Samples retained in sealed vessels
Evil smelling	Results checked against specification
Difficult to wash out samples from flow cells	Out of specification samples repeated

7.1.2 Quality control of perfumes, fragrances and pigments

The Ismatec ASA Aroma system was introduced for the automatic quality control of industrially manufactured foodstuffs and cosmetic additives (fragrances, perfumes and pigments). The system makes it possible to determine physical and chemical properties such as density, refractive index, optical rotation and colour. Depending on the system design, 50 to 200 samples may be examined per batch, in accordance with GLP guidelines. The advantages of the Ismatec ASA Aroma system are accurate reproducibility of the analyses and an ability to process large numbers of samples in a very short time. It was also designed for continuous 24-hour operation.

The system, which is shown in Fig. 7.3, consists of the sample changer, rinsing units, control computer and the Aroma software. Existing measuring instruments may be used subject to compatibility. The software currently supports the following instruments:

1. Colour: Minolta CT-210
2. Refraction: Dr Kernchen Abbemat
3. Polarity: Dr Kernchen Propol (fitted with a serial port)
4. Density: Paar DMA 48

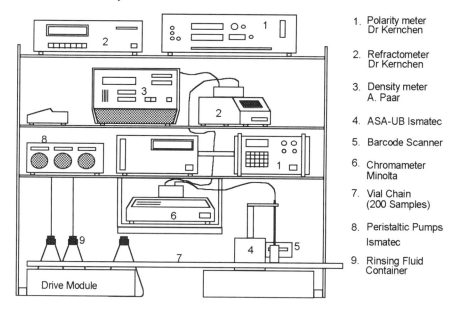

Fig. 7.3 ASA Aroma system for fragrances and flavours designed by Ismatec.
Reproduced with permission of Ismatec UK.

The measuring cycle consists of the following:

a. The user places the sample tubes in the transport unit. If the samples have bar-coded labels they are read at the sampling position with the optional bar-code reader.

b. The sample probe enters the sample and the sample pump starts, sucking the sample through the optical measuring cells. The minimum sample volume for a system with four measuring instruments is 5.5 ml.

Sec. 7.1] Automatic specific gravity and refractive index measurement 199

c. As soon as the cells are full, the sample pump stops. To prevent continued sample movement during the measurement cycle, a pressure-compensation valve opens for a short period. After this, a period of time is allowed for temperature adjustment in the measuring cells.

d. The measuring instruments now read the sample values and send them to the computer. The computer files the sample readings, together with the bar-code or sample batch number, and stores them.

e. The sample is pumped back to the original tube or to waste (switchable). The system is flushed with two different solvents (ethanol or acetone) separated by air segments with the two Mini-S rinse pumps. The contaminated solvents may be collected separately into rinse bottles A2 and A3. The flush times can be changed in the menu so that measuring error through contamination is minimized. It is also possible to preset different times, according to the bar-code.

f. After the solvent wash, another valve is opened so that compressed air can blow out remaining solvent residues (a compressed air supply is essential).

The sample cycle is now finished and the sample transport steps to the next sample. The average sample cycle time is 3 minutes with four measuring instruments.

A block diagram of the system, set up for four measurements, is shown in Fig. 7.4.

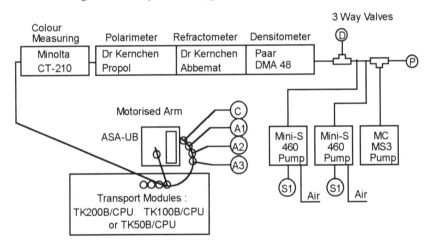

S1 - Cleaning sol. 1 (Isopropanol/H_2O)
S2 - Cleaning sol. 2 (Acetone)
A1 - Waste Isopropanol/H_2O & needle cleaning
A2 - Waste Acetone & system drying
A3 - Waste for probe
C - Calibration measuring instrument (Dest. H_2O)
D - Pressure compensation against air
P - Compressed air connection

Fig. 7.4 Block diagram showing the ASA Aroma system. *Reproduced with permission of Ismatec UK.*

The Aroma software controls the sample change, pumps, valves and connected measuring instruments. The software also receives all measured results from the instruments, filters, stores and prints them.

All functions are menu-driven. Parameters used for the running cycle (for example rinse times) may be changed by using a mouse or a keyboard; changed parameters can be stored permanently. There is no need for the user to have any knowledge of computer programming.

This instrument [1] is well designed and uses somewhat different principles to those described previously. Essentially it uses the design company's expertise and applies it to readily available instruments for measurement of the appropriate test parameters.

7.1.3 Modular analysis system

A different system has been introduced by Index Instruments [2]. This modular analysis system (MAS) provides the analysis of up to four different parameters: refractive index, density, colour and optical rotation. A wide range of viscosities can be handled. The analysis, wash and drying times can be selected by computer control so that cross-contamination can be reduced to undetectable levels. Samples are only in contact with inert materials such as glass, synthetic sapphire, PTFE and 316 stainless steel.

The schematic arrangement of the MAS system is shown in Fig. 7.5. The autosampler will handle up to 40 samples which, when the parameters are defined, can be left to operate under computer control. The MAS system is configured complete to fit user requirements and includes both hardware and software facilities.

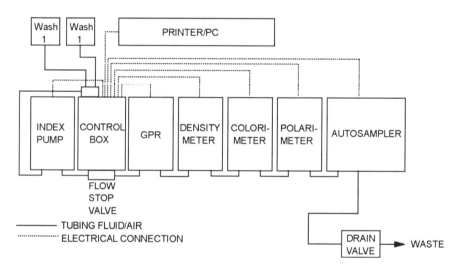

Fig. 7.5   Schematic layout of the Index Modular Analysis System (MAS). *Reproduced with permission of Index Instruments Ltd.*

7.2   COMPUTER-ASSISTED ANALYSIS OF METALS

For the routine determination of analytes in the quality control of the production of speciality chemicals, a combination of direct current plasma emission spectroscopy (DCP-OES) with flow injection analysis (FIA) has been used. Results obtained for the determination of boron, copper, molybdenum, tungsten and zinc in non-aqueous solutions have been published by Brennan and Svehla [3]. The principle has been extended to other analytes, carrier liquids, and solvents, and the details of a fully automatic system have been described by Brennan *et al.* [4].

The combination of FIA with DCP-OES means that constant nebulization is maintained over a long period of time. The introduction of the sample as a plug into the carrier stream causes a transient signal (peak) in the response, which soon decreases to the background level caused by the carrier liquid [5]. This signal is almost as large as continuous responses obtained when large volumes of samples are nebulized over a substantial period of time. In a typical application, signals of 93% of the continuous signal have been produced, with a relative standard deviation of 1.6% over 20 injections.

The instrumentation used for this application consists of a Beckman Spectraspan 4 direct current plasma optical emission spectrophotometer. The DC plasma source is well suited for excitation of emission spectra for a wide variety of elements. The plasma jet consists of three electrodes arranged in an inverted 'Y' configuration. A graphite anode is located in each arm of the 'Y' and a tungsten cathode at the inverted base. Argon flows from the two anode blocks towards the cathode. The plasma jet is formed by bringing the cathode momentarily into contact with the anodes. Ionization of the argon occurs, and a current of about 15 A develops, which generates additional ions which sustain the current indefinitely. The plasma source can easily accept liquid samples. The sample is aspirated into the area between the two arms of the 'Y', where it is atomized, excited, viewed and measured. The plasma temperature is around $5500^{\circ}K$ [6]. This temperature is capable of removing most matrix interferences and chemical effects and improves the sensitivity, thus making the DCP source an extremely versatile tool for a wider range of samples and solvents. The higher thermal energy offers a 10-fold increase in linear calibration range.

The instrument has a Czerny-Turner type monochromator, consisting of a prism and an echelle grating, which projects the spectrum onto the exit slit. For the wavelength range of 200 - 800 nm, reciprocal linear dispersions vary within 0.062 - 0.25 nm/mm. This high resolution virtually eliminates all errors originating from spectral overlap, offering a precision and accuracy (better than would be obtained by flame-emission methods), and improved detection limits for elements which are difficult to excite. The high resolution also permits the use of fairly wide slit heights and widths (for example 500 x 200 μm), resulting in relatively strong signals in the photomultiplier.

7.2.1 Configuration of the FIA/DCP/OES system

In the combined FIA/DCP/OES system, the liquid sample containing the element of interest is injected into a continuous stream of carrier liquid and is transported to the plasma jet for atomization. On its way to the detector, the sample plug is mixed with the carrier liquid and is partially dispersed. The degree of dispersion depends on the distance of the injection point from the plasma, on the volume of the sample, on the flow rate and on the inner diameter of the tubing. These parameters have to be optimized for particular samples, solvents and carrier liquids; the viscosities of the latter two also have major roles. Experimental conditions have to be kept constant for both samples and standards. The sample concentration can be evaluated against properly made-up standards, which are injected in the same way as the samples. Flow-injection systems are characterized by short response times; analytical signals are obtained in between 2 and 5 s, so a high sample throughput is possible.

A schematic diagram of the automated FIA/DCP/OES system is shown in Fig. 7.6. The carrier liquid is transported using a peristaltic pump, at a rate predetermined to suit the particular analysis. Teflon tubing (internal diameter 0.8 mm) is used where appropriate; however, silicone tubing is applied at the roller heads to accommodate both organic and inorganic solvents. Samples are introduced by the autosampler through a loop. The size of the loop is variable; for most applications a 600 µl loop is sufficient.

A precision-controlled flow-injection valve is necessary to keep the dispersion to a minimum and to keep the volume and length of the sample plug as low as possible. A Teflon six-port valve was found to be best for the purpose. A schematic flow diagram of the valve configuration for filling and injecting is shown in Fig. 7.7. Two ports are used for the loop; one is shown for the inlet and one for the outlet of the carrier stream. The fifth port is used for carrying the plug to the DCP/OES detector, and the sixth port is used as an excess drain. This valve can be controlled manually through an electric switch or by a microcomputer. The intelligent autosampler is also controlled by a microcomputer.

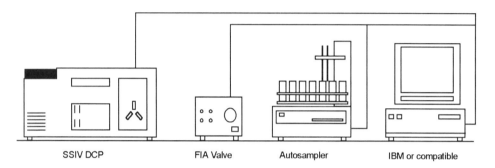

Fig. 7.6   Schematic diagram of the automated FIA/DCP/OES analyser.

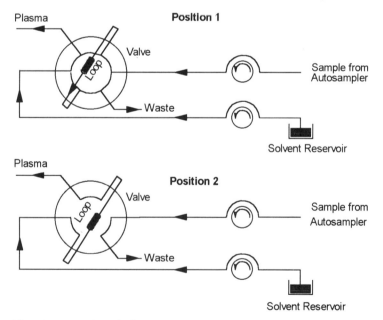

Fig. 7.7   Schematic diagram of valve configuration.

The FIA valve is connected after the pump, approximately 25 cm from the nebulizer to minimize dispersion (this is the shortest distance attainable between the valve and the DCP jet). When attempting a new procedure, a 15 s wash-out period should be allowed between injections; this was found to be adequate to minimize any 'memory effects'. However, the time can be shortened to 1 - 2 s for most applications, once the optimal experimental conditions have been established.

All the experimental parameters can be stored by the microcomputer, and all operations can be controlled by it.

7.2.2 Signal acquisition and the data management system

This system consists of a microcomputer (ESR class) with an RS232 interface, an enhanced graphics adaptor screen, printer and a chart recorder.

The software (available from P.S. Analytical) data acquisition program is designed to operate with an IBM, or compatible, computer working under MS DOS or a similar operating system. The program is user friendly and gives a step-by-step guide through its facilities. It can collect, store and process data generated as line intensities in the spectrometer. Data is transferred through the serial asynchronous communications interface continuously. When measurements for standards are completed (by standard addition or standard curve), a curve is automatically displayed on the screen to suit the type of analysis. The program may be interrupted at any time for logging additional data or information. The data collection system allows statistical analyses to be carried out on a series of similar measurements of different samples, calculating means and standard deviations. A similar option is available for the statistical analysis of repetitive measurements on the same sample.

The technique developed allows rapid and routine determination, both at trace and higher concentration levels, in either aqueous or non-aqueous solvents, of approximately 70 elements in the periodic table. It is fully automatic with all the operations, including sampling and clean-out of the plasma jet, controlled by a microcomputer. Being able to control the time allowed for clean-out of the plasma jet is especially useful in the determination of elements which tend to cause 'memory effects' in later samples.

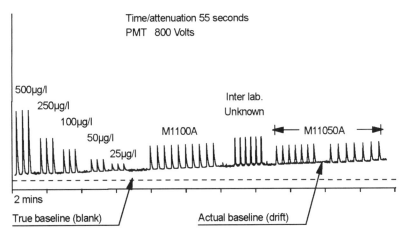

Fig. 7.8   Reproducibility of signals with a shifting baseline.

Thus, for the determination of difficult elements, such as tungsten, boron and molybdenum, this technique produces substantially more precise and accurate results than traditional batch operation. The DCP/OES parameters can be set to various levels to accommodate wide concentration ranges.

The flow-injection method is extremely valuable in correcting for baseline drifts (which originate from uncontrollable thermal and electronic changes during the course of operations). This drift can cause high errors, especially at trace levels, when the conventional continuous nebulization technique is used. The elaborate and time-consuming correction procedures in the conventional technique are not necessary with the FIA method - the baseline is defined by the emission obtained from the carrier liquid, and is reproduced between each sample injection, see Fig. 7.8. The peak heights, when measured from the baseline (as done in the FIA technique), are very reproducible; if, however, these peaks were measured from the initial blank value (called the 'true' baseline in Fig. 7.8), substantial errors would occur.

Automation is especially advantageous for analysing large numbers of samples on a routine basis. The flow-injection method requires low sample volumes, but even the recommended 600 μl loop size can be reduced to approximately 100 μl without substantial losses in sensitivity, accuracy or precision. In certain applications involving ICPs samples as small as 20 μl have been reported [7]. There is little doubt that sample introduction with a flow-injection valve and driven by a peristaltic pump or another solvent delivery pump is superior to other micro-sampling techniques [8]. Other advantages of flow-injection analysis include the potential for providing a complete analytical calibration curve from a single standard solution [9]; the possibility of using the equipment with other instrumental analytical techniques (ICP/OES, atomic absorption spectrometry, UV/visible molecular absorption spectrometry, spectrofluorimetry and voltammetry); and the relatively low cost of the instruments and the operators' time.

Flow-injection methods are quicker, more precise and use less reagents than other techniques; in addition, they are very useful where only a limited amount of sample is available. The main advantages of the combination of FIA with DCP/OES are the increase in precision in sample handling, fewer physical interferences, a higher throughput of samples and versatility towards physical and chemical properties of reagents. Some minor disadvantages are the loss of sensitivity compared with continuous nebulization, and the fact that only one element can be determined at a time.

The system described here was developed for use in the research and development area but has been readily transferred to a quality-assurance environment. The ruggedness of the DC Plasma Spectrometer greatly extends the usefulness of the modules and drastically reduces sample preparation procedures, which are time-consuming and error-prone.

7.3 AUTOMATED TABLET DISSOLUTION SYSTEMS

In the pharmaceutical industry, it is important that all products are properly tested and validated prior to release for sale. One of the most important tests is the determination of tablet dissolution rate. In this procedure tablets are immersed in a suitable medium to mimic the action of the stomach and the release of the active ingredients monitored over a period of time. Automation of these procedures is obviously important for various reasons including cost, accuracy of analysis and for validation of the results according to good laboratory practice.

Analysis is normally carried out in a series of thermostatically controlled baths according to standard conditions. Automation is required to control the instruments in the system, select the samples, analyse the sample withdrawn, collect and analyse the data, and, finally, to present the data produced, in a readily digestible form, for the laboratory itself, the managers, and also for the legislative bodies involved. Typically, thermostatic baths with six or eight analysis batches and a control standard are required. Analytical data is usually provided by ultraviolet spectroscopy at single or multiple wavelengths. With the availability of diode array spectrometers more systems are now being installed with wavelength-scanning over regions of interest.

A further procedure involves the inclusion of an HPLC column separation analysis. However, the rate of sample uptake must be controlled by the analysis time on an HPLC column.

Figures 7.9(a) and 7.9(b) show two different approaches to the sampling of the dissolution baths prior to analysis. Figure 7.9(a) shows the most commonly installed system, which uses a series of spectroscopic cells, one to coincide with the bath being sampled. The system uses a peristaltic pump to withdraw the sample through the pump tubing and into the flow cell which is moved in co-ordination into the light path of the spectrometer for analysis. The system shown in Fig. 7.9(b) was developed in association with Glaxo Operations, Barnard Castle, County Durham, UK, and has been described by Lloyd *et al.* [10].

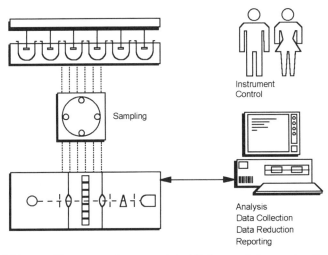

Fig. 7.9(a)   The most popular commonly installed approach for sampling dissolution baths using a series of flow cells.

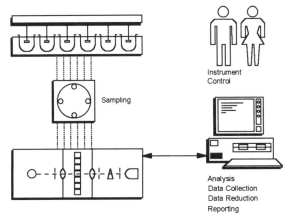

Fig. 7.9(b)   Approach developed for dissolution sampling using only a single flow cell.

The second system is preferable in that all aspects of the analysis are controlled by a computer. The system can easily be integrated with any detection system according to the analysis being monitored. A patented valve-switching system, which can be randomly addressed under computer control, selects the samples. These are withdrawn through a single cell and sent to waste using the peristaltic pump. Because a single cell is used, there are errors from cell variation. The sample is only in contact with pump tubing when it is being thrown to waste. Forward and reverse flows are available with the pump system so that any filtrate on the filter prior to the cell can be removed before sampling. A simple and neat hydraulic system is provided, and any problems are easily diagnosed.

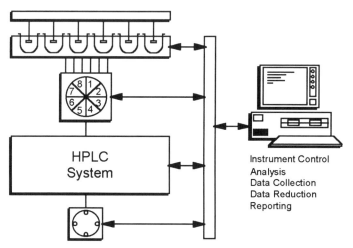

Fig. 7.10   Automated dissolution analysis using HPLC detection.

When a wide range of analyses is required, the first system requires a considerable investment in the supply of the six and eight cells, according to the number of analyses required. The second approach uses only a single cell for each analysis. A minor disadvantage of the second approach is that it requires a separate valve to add medium to replace the sample. However, the system developed by Glaxo takes the loss of sample into account and complete computer control is provided using an HP85-based computer system. Figure 7.10 shows a similar system where an HPLC system is used to replace the UV detection system.

Table 7.2 shows the control computer's operation in the Glaxo system. It can be seen that all aspects of the system can be initiated and tested before starting the analysis. The system is modular and can easily be configured according to users' preferences. Data can be displayed in either a tabular format or as a graphical display, to suit user requirements.

Although the Glaxo system has many advantages, it has not been an enormous commercial success. This is partly because it has not been adopted by a large instrument company, and most importantly, the software was designed to operate using Hewlett-Packard HP85 and subsequent models, which are not IBM compatible - the IBM compatible computers have become the accepted industry standard. When the control software and data collection routines are made IBM compatible, then the single-cell system will probably become more popular.

Table 7.2   Computer control operation

7.4   ROUTINE DETERMINATION OF MERCURY AT LOW LEVELS

Mercury is used in many industries, but recently there has been increasing concern about its effect on the environment. This, together with legislative pressure, has dictated the need for analytical measurements at lower and lower levels.

Godden and Stockwell [11] have described a specific fluorescence detection system to provide fully supported analytical systems for routine analysis of mercury at low levels. The fluorescence approach provides a wide linear dynamic range and extremely low detection limits. P.S. Analytical's Merlin Plus System provides a fully automated system which will produce results at a rate of around 40 per hour. This is due to the optimization of the optical design of the detector, coupled to the inherent features of the fluorescence technique.

Figure 7.11 shows a specific configuration of a fully automatic system designed for laboratory use. The version illustrated has evolved from previous designs and feedback from existing users. The preferred configuration consists of an autosampler (this holds cups with 27 ml capacity, which is sufficient for repeat analyses) linked to the vapour-generator system. The mercury vapour produced is transferred into the fluorescence detector. The interface between the vapour generator and the Merlin provides a flow of mercury vapour in an argon stream which can also be sheathed in a further flow of argon. The system is also briefly described in Chapter 5.

The interface provides efficient transfer of samples into the Merlin, and, most importantly, a rapid flush-out; there is no hold up of mercury (which is a feature of the commonly used atomic absorption techniques). To aid the transfer of mercury vapour, the tin(II) chloride regime is used, together with a gas/liquid separator designed for this task. Mercury is sparged from the reaction vessel into the Merlin Detector. Full automation is provided by using a simple standard DIO card fitted into an IBM compatible computer system with the PSA TouchStone© software. This is an easy-to-use menu-driven system which controls the modules used in the instrumentation, calibrates the system, collects, collates and reprints the results, and which links to host computer systems.

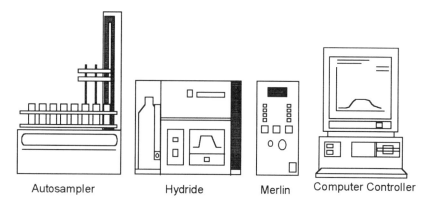

Fig. 7.11   Configuration of fully automated mercury analyser.

The methodology has been specifically developed in conjunction with Yorkshire Water Laboratory Services, one of the primary users of the instrumentation.

Generally speaking, current legislation requires the measurement of total mercury. Samples held in containers on the tray are fed, under computer control, by continuous-flow techniques into the vapour generator. As the mercury evolves, it is transferred to the detector and the measurements are made. The chemistry of this reaction and the design of the gas-liquid separator have been the subject of considerable investigation. Tin(II) chloride in an acid medium has been found to be the most flexible and valuable reagent for the analysis of routine samples. Essentially, tin(II) chloride will react with divalent mercury in the sample. Any organic mercury present in the sample must be converted into the inorganic form. Many procedures are available for this, for example acidified potassium permanganate or dichromate. The inherent sensitivity of the fluorescence procedure shows up the problems in these chemical processes, especially with high blank levels. Yorkshire Water has developed a method based on the use of a potassium bromate/bromide reagent which has little blank interference, and the reaction is extremely quick.

Consideration was given to automating the bromate/bromide reaction in series with the mercury vapour generation. This was rejected for two reasons: firstly, the addition of another process complicates the design of the manifold and increases the possibilities of errors. Secondly, the reaction with bromate/bromide is a single-step process that requires only time to activate and complete - 45 minutes is ideal. So if the analyst prepares a tray of samples at the end of the day and leaves these overnight, in this case the term 'A' in the cost equation is very low and the cost of this is minimal (Chapter 1). When the analyst starts the analysis next morning he or she simply sets up the system, loads the first tray for analysis and then commences. Whilst the first tray is being analysed, the second tray can be digested to convert the organic mercury into the divalent form.

Several analytical procedures have been developed using this chemical regime for rivers, effluents, urine and some organic-type materials. A single regime provides reliable results and avoids a cluttered and complicated manifold design. It clearly identifies what the automatic system is designed to achieve and allows the transfer of this technology to other laboratories easily and reliably. Treating the samples with bromate/bromide reagent at the actual sampling point obviates time delays before the determination of total mercury.

7.4.1 Sample collection and preservation

Samples are taken in 300 ml disposable paper cups after pre-rinsing with the sample twice. An aliquot of the sample is then immediately transferred from the paper cup to a clean 100 ml measuring cylinder containing 15 ml of 33% v/v hydrochloric acid and 2 ml of 0.1N $KBrO_3/KBr$. The cylinder should be filled to a volume between 90 and 100 ml. The total occupied volume should then be recorded on the cylinder label, together with the sampling point details. If a filtered mercury sample is required, the filtration should be carried out as soon as possible after sampling. The filtrate is treated as above. The sample is then allowed to stand for at least an hour to ensure complete breakdown of organomercury compounds. If a distinct bromine colour remains, the sample may be stored for at least two days if mercury contamination is avoided.

If the sample contains a lot of organic matter and no free bromine remains a further 2 ml of the potassium bromate/bromide reagent should be added. A blank containing twice the normal amount of this reagent should also then be prepared.

7.4.2 Operation

Reagents and standards are prepared as above. The instrumentation is set up according to the manufacturer's recommendations and allowed to stabilize for 10 minutes. A quick check with coloured solutions will ensure that the equipment is functioning correctly.

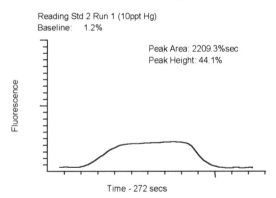

Fig. 7.12   Typical signal response for 10 ppt of standard mercury. The peak shape is consistent with complete digestion and can be used as an indicator of interferents.

The reagents, blank and samples are then presented to the vapour generator. Argon is the preferred transfer gas, since nitrogen and air quench the fluorescence and reduce sensitivity by 8 and 30 times respectively.

The instrumentation is then calibrated over the range of interest, using the TouchStone© software to prompt and store the results. A typical signal response is shown in Fig. 7.12. This represents the output obtained from 0.010 ppb of mercury in water. A calibration graph computed by the method of least squares shows excellent correlation, and confirms the linearity and sensitivity of the method and the applicability of atomic fluorescence. Where blanks are matched closely to the samples, the graphs pass through the origin and the equal weight slope average method of calculation is valid according to theory. For real world samples it is essential to analyse a reagent blank with the calibration, fit type is based on the method of least squares.

7.4.3 Control and data collection

Two features of the complete system are worth stressing: the autosampler design and its control and implementation with the specific software. The TouchStone© software was designed by specialists in instrumentation with a great deal of experience of analytical chemistry. It uses computer techniques to give the analyst what is required, and it achieves its objectives easily and reliably. It is also easily understood by people who are not computer specialists.

The autosampler used is driven by a stepper-motor controller, which can address any sequence on the tray. Any number of sample positions and bottle types can be placed around the perimeter in one or two circular rings. The TouchStone© software has in its design a tray pattern based on tray position and the number of steps from a reference point that the tray has to be driven to reach the position. Using the RS232 link, the sampler can be moved between one position and another. Switching between the inside and outside of the tray rings involves the use of an inert switching valve. A simple 0 or 1 command by the computer program activates or deactivates the valve and selects the sample position from which the sample is withdrawn.

The hydride generator is controlled by a DIO card in the computer and this triggers the measurements from the fluorescence detector, which are recorded by a BCD interface.

As the pressure for the validation of analytical results to be validated increases, the computer software and the system have had to be modified so that they can operate in a 'hands-off fashion' to a set protocol design. The TouchStone© software provides this with the 'Program' facility. Procedures and instruments can be changed within a single run.

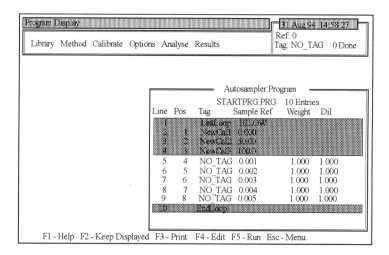

Fig. 7.13   Computer display of the program mode showing the inherent flexibility available to the user.

The program option allows the user to run a batch of samples automatically, changing between Tag and Single numbers. Alternatively, the user can run a batch of samples with automatic reslopes, checks or even a completely automatic calibrated analysis with re-checks and re-slopes. This feature is easily understood by reference to Fig. 7.13.

Where a check is required, the necessary precision can be defined. Should the result fall outside the preset limit then the analyst can set up the action to be taken by the computer. Three options are provided: A = abort, I = ignore and R = repeat. When a repeat is requested, the analyst can also identify the line number in the program where the sampler should be directed to re-start the analyses. The software routines also allow the analyst to edit multiple runs of data. Additional modules and features have been added to the software routines to substantially extend the flexibility of the instrumentation.

7.4.4 Mercury in seawater at ultra-trace levels

To extend the levels of detection for mercury still lower, several workers, especially in this area of atomic absorption techniques, have chosen to collect the mercury on gold or other noble metal trapping systems prior to revaporizing the mercury into the measurement technique. Figure 7.14 shows the configuration of a specific system to concentrate mercury onto an amalgam preconcentrator prior to analysis.

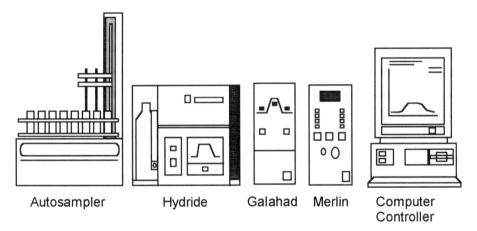

Fig. 7.14   A specific system for mercury concentration.

The instrumentation adds this device on-line with the standard units for mercury analysis and is also controlled by the software. Steps have been taken to minimize the blank levels of mercury in the solutions, gases and reagents.

Basically, these involve the addition of three electronically activated Teflon valves to direct the flow of gas from the gas-liquid separator and over the Galahad instrument. One valve is controlled directly from the vapour generator and this only diverts the argon containing mercury stream over the gold trap for a preset time. The additional two valves are controlled by the Galahad cycle so that the revaporized mercury can be directed to the Merlin detector for measurement. In this manner the flows for preconcentration and the flows for measurement can be optimized individually. The operation of these valves is shown schematically in Fig. 7.15.

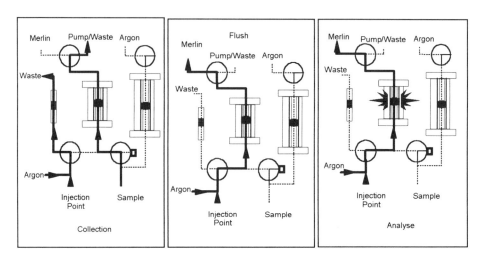

Fig. 7.15   Schematic layout of mercury detection system showing valve sequencing for collection, flush and measurement.

Cossa [12] has used this approach to monitor the levels of mercury in seawater samples.

Figure 7.16 shows a typical calibration graph for the range of 0 - 2.5 ppt. Analysis rates of around 10 - 15 samples per hour are obtained at the sub ppt levels.

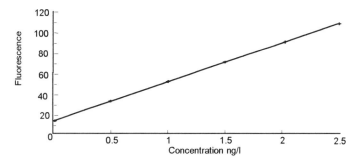

Linear Correlation Coefficient = 0.9985

Fig. 7.16   Calibration graph of vapour generator linked to gold amalgam preconcentration system.

This method combines the strategy of continuous flow and batch operation. Whilst it complicates the analysis, it can be shown to have two advantages. It can further reduce the detection levels obtained. Figure 7.17 shows the blank response and also a blank spiked with 0.5 ppt of mercury. A calibration graph produced by the method of standard additions shows how the detection levels are reduced. The second advantage is not necessarily in the concentration of the mercury, but more the removal of interference effects. Organic vapours may absorb in the 254 nm region and should they be present they can effectively reduce the fluorescence response. The analyst has two alternatives to overcome this, either the removal of all of the organic material by sample preparation techniques, or absorption of the mercury onto gold and removal of the interferents in this way. For some analyses the second option is preferred.

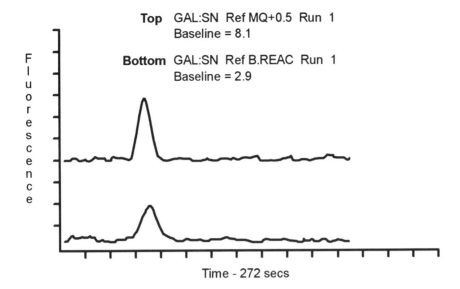

Fig. 7.17 Atomic fluorescence detector response from samples preconcentrated on gold traps prior to revaporization.

However, it must be stated that the disadvantage of such techniques is the additional complication of two procedures, and also a significant problem relating to background levels. These, of course, will be concentrated by the technique. Procedures to reduce these background levels can place considerable constraints on the analysis and also significantly increase the cost of the instrumentation involved. Given adequate attention to detail, the availability of mercury-free reagents, and a clean working environment, the combination of techniques described here can lower the detection by a further order of magnitude.

7.4.5 On-line analysis for low-level mercury determinations

Since the introduction of the first fully automated laboratory-based mercury analysers in 1989, a number of other systems have been introduced. Around 20 different systems are available with varying levels of performance, and many claim to measure mercury at low levels. The analyst can therefore be forgiven if the determination of mercury is considered a trivial problem. Despite the various claims, the determination of low levels of mercury is very difficult to achieve. By far the most significant problem is the collection and presentation in the laboratory of a fully representative sample. Mercury levels can be reduced or magnified by several orders of magnitude if proper control is lacking. The use of on-line methods of analysis would serve to overcome many of these problems.

The methods for mercury analysis described earlier are based on continuous-flow technology and although this is easy to translate these procedures into an on-line regine, several problems are associated with this. These are (a) presentation of a representative sample; (b) conversion of all forms of mercury in the sample into the divalent form prior to reduction to mercury; and (c) engineering the system to include sufficient robustness, control and flexibility.

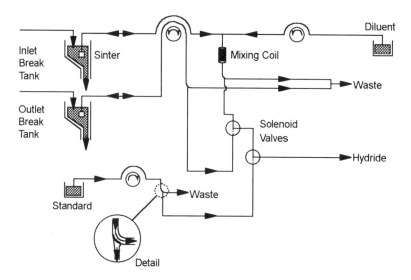

Fig. 7.18   Schematic representation of manifold and sampling input.

Figure 7.18 shows the schematic arrangement for an on-line system incorporating adequate filtering of the sample stream to prevent high solids from entering the flowing stream. A valve manifold arrangement is also incorporated, such that samples and standards can be analysed under computer control. In the arrangement shown, two different samples streams can be analysed in a set sequence and high and low standards used to check for any instrumental drift.

Once a representative sample of the matrix is available, the second most difficult problem is to provide adequate chemical procedures to convert all of the mercury species into inorganic mercury. The chemistry required to complete this will be dictated primarily by the nature of the sample. The three different procedures outlined in Fig. 7.19 (a-c) are designed to cope with a range of chemical situations.

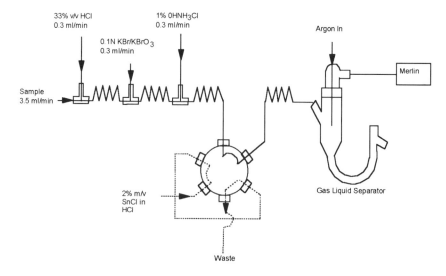

Fig. 7.18(a)   Chemical regime using bromination.

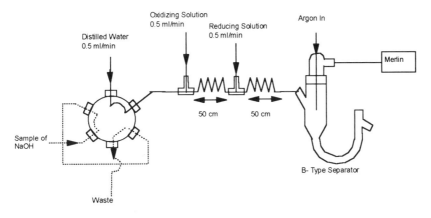

Fig.7.19(b)   Chemical regime for caustic soda.

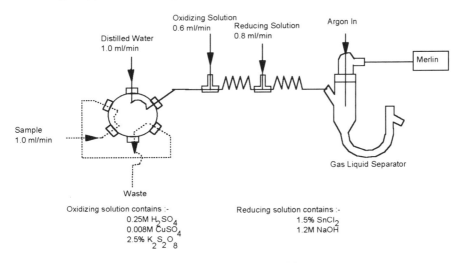

Fig. 7.19(c)   Chemical regime using catalysed persulphate

The bromination procedure mirrors the techniques used in the laboratory environment. An alkaline permanganate procedure has been found suitable for very alkaline applications, especially in chlor-alkali plants. The copper sulphate/potassium persulphate procedures have been successfully applied to high-pressure liquid chromatographic applications. Each of these chemical regimes has merit and can be tailored to fit the actual sample matrix being tested. For on-line applications a further order of reliability is required as instruments will be operated for several months. Slow sampling speeds are of prime importance in order to limit consumption of reagents. Equally, the instrument must be simple to operate and require little, if any, maintenance. Reagents should be changed not more often than weekly.

Figure 7.20 shows a schematic layout of the on-line system. All materials of construction are either 316 stainless steel or an inert polymeric material such as Teflon. Reagents are pumped at 0.3 ml/min, and the six-port change-over Teflon rotary valve ensures that only small volumes of samples or reagents are injected into the flowing stream for analysis. With the system configured as described above, it is also possible to define a flexible sequence of analyses so that the data fully represents the current status of the instrumentation, and the mercury levels are correctly monitoring the process being controlled. Results can be directly displayed in the required format to the main process control room via a D/A card included in the instrumentation.

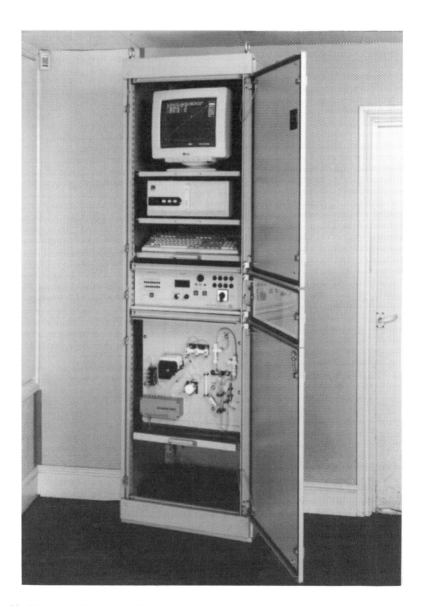

Fig. 7.20   Photograph of on-line system.

The system provides discrete analytical measurements on a fixed-time cycle and does not sense continuous levels. This facility enables reagent requirements to be minimized, making a subsequent reduction in servicing requirements, i.e. reagent levels can be refilled at weekly intervals. Precision levels are in the region of 2 to 3% in routine applications.

7.5 AUTOMATED pH AND CONDUCTIVITY MEASUREMENT

Measurement of pH and conductivity for a range of aqueous samples is common in most water laboratories. Surprisingly, many laboratories still use manual methods for these measurements. The automation of these techniques is not easy. Two approaches to solve this automation problem, described in this section, both used the same large volume autosampler.

It is difficult to obtain pH and conductivity sensors from the same manufacturers, and also to find systems that will operate without mutual interaction. When the author was asked to configure such an automated system, a suitable commercially available sensing system was sought. Solomat Manufacturing Ltd (2 St Augustine's Parade, Bristol, UK), has a complete range of pollution monitoring equipment designed for on-site testing. The company also seemed to have suitable pH and conductivity sensors, appropriate hardware to make measurements and also to link to IBM compatible software for data collection.

When the system which was to be linked to the autosampler system was purchased, it was found that the conductivity cell was excessively large and primarily designed to be fitted to a ½" pipeline; for analytical purposes this was clearly too large. However, the pH cell provided an Ingold standard electrode system with dimensions which allowed it to be placed within the conductivity sensor, as shown in Fig. 7.21. It is therefore possible to configure a flow cell in which the conductivity sensor houses the pH electrode. The latter also serves to take up most of the cell volume so that the wash-out requirements are not too excessive. The cell is filled by taking a sample from the autosampler with pump 1; liquid fills the cell and overflows via the weir arrangement and the top of the cell.

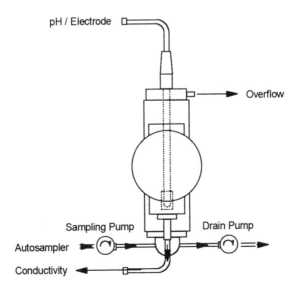

Fig. 7.21   Schematic diagram of automated pH conductivity instrument, showing the sample transport mechanism.

Stopping this pump allows the solution to stabilize before measurement. After measurement, activation of pump 2 (which runs at a faster flow-rate), cleans the cell of liquid. Two flushes of a sample were found to give a representative measurement even when changing from high to low conductivity in the sample line. The pumps, in this instance, act as a series of valves to wash the cell, set up a representative sample, take the measurement and then wash the cell prior to the next sample.

The autosampler system is controlled by an IBM computer system, as is the series of pumps for the cell and the sample wash pot. An Archer single board computer (Sherwood Data System) programmed via ASCII strings along an RS232 interface, controls the pumps and the autosampler, setting up a stable representative sample which is then measured by using the standard Solomat software and hardware.

Programs have been written around the existing Solomat software to establish the correct protocols and to pass control from the autosampler pump system and back to the measurement system at the appropriate time. Additional measurements could be integrated with those of pH and conductivity, providing that a suitable cell arrangement could be constructed. The system was shown to function well and gave accurate measurements with a 30 ml sample.

To maximize the usefulness of this approach, however, it would be desirable to reconstruct the data-collection software and to formulate a computer package which can be more easily interfaced. This was not possible for a one-off development and the constraints imposed by the inherent software design had to be tolerated.

Almost at the same time, another company, ChemLab Scientific Products (CSP) [13] in conjunction with Thames Water, designed and developed an alternative system. A schematic arrangement of the unit is shown in Fig. 7.22. ChemLab's new analyser automates four of the most labour intensive procedures left in the analytical laboratory.

The CSP Water Analyser simultaneously measures pH, conductivity, colour and turbidity at a rate of approximately 60 samples/hour. Up to 80 samples can be loaded onto the turntable and the analyser, which is under the control of a PC - no further work is needed. A specification of the analyser is set out in Table 7.3.

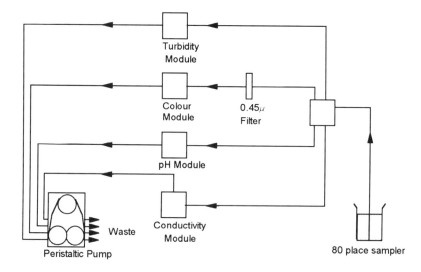

Fig. 7.22 Schematic layout of the ChemLab Scientific Products water analyser designed in collaboration with Thames Water.

Table 7.3   Specification of ChemLab Scientific Products' water analyser

Sampler	80 cups each of 35 ml	
Sampling rate	approximately 60 samples per hour	
Standard measurement ranges	pH	1.00 - 13.00
	conductivity	10 - 1999 µS
	'true' colour	0.5 - 20.0 Hazen units
	turbidity	0.10 - 4.00 NTU
	temperature	+5 - +50°C

This analyser is a computer-controlled automated batch analyser, using a stop-flow principle to analyse for pH, conductivity, turbidity and colour. The principle of analysis for each module is based on the recommended methods as detailed in the 'Examination of Waters and Associated Materials' issued by the Standing Committee of Analysts of the Department of the Environment. The temperature of the sample liquid flow is measured in order that temperature-compensated results of pH and conductivity can be quoted.

A peristaltic pump pulls the sample stream to each of the module detectors for a selected time period: this is the 'Pump Time'. The peristaltic pump then stops and the detectors are allowed to come to equilibrium: this is the 'Measure Time'. At the end of the selected measurement time, the steady-state reading is logged by the computer and the analytical cycle proceeds. If a steady state is not achieved, the result is flagged.

Turbidity is caused by any solid material which is dissolved or suspended in a liquid. The intensity of light scattered by a sample is measured and compared with that measured for standard formazin suspensions, and expressed as nephelometric turbidity units (NTU). Colour is determined as the absorbance (measured spectrophotometrically at 400 nm) of the sample filtered through a 0.45 micron pore size membrane filter [12,13].

Electrical conductivity is a measure of a solution's ability to conduct electricity. The conductivity depends on the concentration of the ions present and the temperature of the solution. A standard conductivity flow cell is integrated into the instrument design.

The instrument is built around a commercially available autosampler designed by P.S. Analytical Ltd, Kent, UK. This has a helical tray and the probe remains static. All the analytical modules are incorporated in to the upper housing of the sampler system. The unit is compact but it is difficult to maintain and service; nevertheless it has been widely accepted as a standard in the UK water industry.

7.6   SUMMARY

The systems described in this chapter are real world examples and illustrate a number of design concepts, but are not necessarily the best and most complete designs available. The solution to a problem depends on the technology, the staff, and the time available.

REFERENCES

[1] Ismatec SA, Feldeggstrasse 6, CH 8152 Glattbrugg, Switzerland.

[2] Index Instruments Ltd, Bury Road Industrial Estate, Ramsey, Huntingdon, Cambridgeshire, PE17 1NA, UK.

[3] Brennan, M.C. and Svehla, G., *Z. Analytical Chemistry*, 1985, 335, 893.

[4] Brennan, M.C., Simons, R.A., Svehla, G. and Stockwell, P.B., *Journal of Automatic Chemistry*, 1990, 12, 183.

[5] Ruzicka, J. and Hansen, E.H., *Flow Injection Analysis (2nd edition)*, Wiley - Interscience, Chichester, UK, 1988.

[6] Skoog, D.A., West, D.M. and Holler, J.M., *Fundamentals of Analytical Chemistry*, (5th edition), Sanders College Publishing, 574.

[7] Faske, A.J., Snable, K.R., Boorn, A.W. and Browner, R.F., *Applied Spectroscopy*, 1985, 39, 542.

[8] Tyson, J.F., *Analyst*, 1988, 110, 419.

[9] Greenfield, S., *Industrial Research and Development*, 1981, 23, 140.

[10] Lloyd, G.R., Tranter, R.L. and Stockwell, P.B., UK Patent Application 8402538, 1984.

[11] Godden, R.G. and Stockwell, P.B., *Journal of Analytical Atomic Spectrometry*

[12] Cossa, D., Personal communication, 1993.

[13] Chemlab Scientific Products Ltd, Construction House, Grenfell Avenue, Hornchurch, Essex RM12 4EH, UK.

[14] Methods for the Examination of Waters and Associated Materials - Colour and Turbidity of Waters 1981 (HMSO).

[15] Methods for the Examination of Waters and Associated Materials - The Measurement of Conductivity and the Laboratory Determination of the pH of Natural, Treated and Waste Water 1981 (HMSO).

8

The future of automatic analysis

8.1 INTRODUCTION

The only certain prediction for the future of laboratory automation is that whatever is seen in the crystal ball at this moment in time is likely to be wildly wrong. One thing is sure, the role of automation will become increasingly important and it is therefore vital that analytical chemists rise to the challenge and take an active part in the definition of the regulatory requirements. The analytical chemist has for too long been the poor relation in the field of chemistry. Although the organic chemists, the biochemists or the physical chemists often get the praise, it is clearly the analytical chemist in the end who is left with the responsibility to provide the data and the measurements. Limits of detection and analytical requirements are often decided by legislators without reference to the analytical chemist, who is the only person who can decide if these limits and procedures are achievable and reliable. Automation has come a long way since the mid-1970s when *Automatic Chemical Analysis* was first written. Today's system designers have often not been through the early days of development and frequently do not have a clear understanding of the analytical chemistry at the heart of the problems they are addressing.

8.2 INFLUENCE OF COMPUTER POWER

Clearly the major difference between the 1990s and earlier times is the availability of inexpensive and reliable computing and control facilities. Time is a fundamental part of laboratory automation and, in the area of computer data processing, the rapid changes that have taken place over the last 25 years are quite startling. Results from analytical measurements can now be almost instantaneous. There is a ready access to computer power, commercial interface cards and even bespoke software.

Speed of response to data collection and reporting has changed significantly since the 1960s when interfacing was not so advanced and data had to be manually transferred to a computer. Figure 8.1 provides a schematic representation of the processes involved in this work. The software had to be written and checked for errors prior to any data being analysed. Often the user was frustrated for days by simple syntax errors or even a tear in the paper tape.

In the 1970s, progress was made along several fronts; for example, modems became available to transfer information over the telephone lines and hence cut out postal delays. In addition, instrumental data could be provided both as magnetic or paper tape output. Of course there were delays occasionally, but generally significant time improvements resulted.

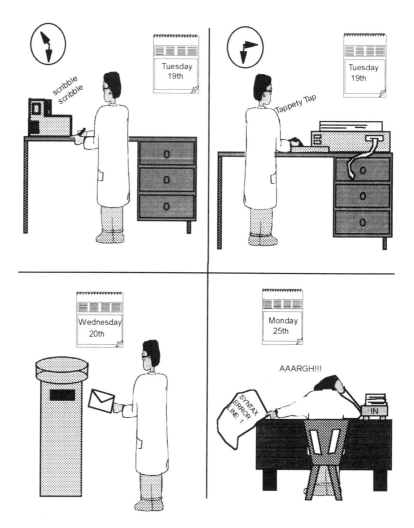

Fig. 8.1   Schematic representation of processes involved in manual data collection and reporting (circa 1960).

In the 1980s, the advent of microprocessors speeded up the processes still further, but efforts had to be made to provide programs to collect and compute data. Today, it is possible to get a data-handling system up and running in a few days and ensure results are computed on-line. In the area of computing, then, the time element has been greatly reduced.

There is no doubt that further improvements in computing power will provide significant changes in laboratory automation. Access to computing power will make individual instrumental modules more flexible and able to meet changing needs rapidly.

While there have been many significant changes in the availability, cost and reliability of computer hardware, the main hold-up to future progress may well be software development timescales. Currently, there are a number of flexible data collection and reporting software packages readily available; the most notable, perhaps, being the National Instruments range of products [1]. These, along with the available interfaces, will enable the systems designer quickly to configure instrumentation to fulfil automation needs. Future developments in software engineering to enable analysts to translate their objectives into real systems would obviously be beneficial. In particular there is an urgency for the analyst to be able to communicate the requirements to the software engineers and for the required computer software to be available economically and within tight time scales. At present, interfacing to commercial instruments which requires software revision is often out of the question because of the cost penalties involved.

Settle [2] has described the evolution of analytical instrumentation over five generations. First-generation instruments consisted of simple devices such as burettes and balances, in which the analyst obtained data, point by point, through manual and visual interaction with the device, Fig. 8.2. After manual manipulations, analysts recorded values associated with visually significant effects, such as colorimetric end-points. The analyst was totally involved in all aspects of determination.

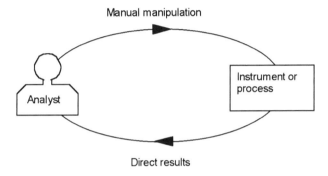

Fig. 8.2   First-generation systems.

From the late 1930s to the mid-1960s, the second generation appeared. These instruments used sensors to convert chemical or physical properties into electrical signals. Electrical circuits were then employed to produce meaningful data (Fig. 8.3). Analysts interacted with these instruments using knobs and switches to obtain data from output devices such as analogue meters and strip chart recorders.

Digital displays began to replace analogue meters in the late 1950s and improvements were made in the conversion of analogue signals into digital output.

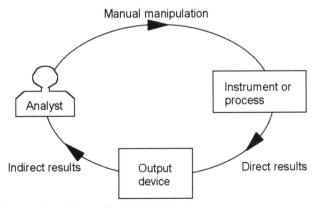

Fig. 8.3   Second-generation systems.

The late 1960s saw the appearance of dedicated laboratory mini-computers and the third generation of instrument systems (Fig. 8.4). The computers were interfaced to existing instruments and were used primarily to log and process data. In some cases, simple instructions could be sent to the instrument by programs resident in the mini-computer. It was also possible for the computer to optimize instrument conditions in real time by monitoring output data.

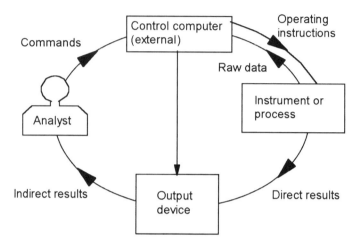

Fig 8.4   Third-generation systems.

However, in these systems, the instrument remained an independent component that could function without the computer. Analysts had to become very familiar with the computer in order to use these systems.

The fourth generation of instrument systems (Fig. 8.5), which emerged in the late 1970s, were the direct result of microprocessor technology. Computers became an integral component of the instrument, receiving input from the analyst, controlling the measurement tasks, processing data and outputting it in many forms. These systems were also capable of running diagnostic programs to check the operation of system components.

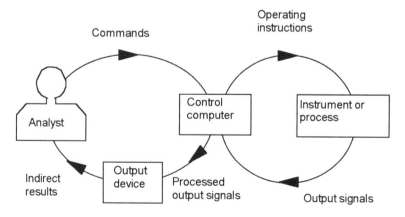

Fig. 8.5   Fourth-generation systems.

Increases in processor speeds and storage capacity allowed these system to acquire and process data rapidly. Many fourth-generation systems became nodes in laboratory computer LIMS networks. They communicate with host computers to receive instructions for analyses and for transferring results. Programs and values of parameters for specific analytical methods can be stored in memory and recalled by the analyst as needed. While the analyst found interaction with these systems easier, he or she became further removed from the system components and often more dependent on the vendor's software. Tailoring requirements to individual user requirements was often not viable with this approach.

The fifth-generation systems currently available are a result of the continuing improvement in the price/performance ratio of computers. The integration of instruments for measurement with apparatus for sample preparation and transfer enables these systems to perform all of the functions required for intelligent automation of methods. In previous generations, the tasks associated with sample preparation and transfer were performed manually, or, in a limited number of cases, carried out by semi-automatic devices, such as autosamplers, linked to an instrument.

The efficient user interfaces of fifth-generation systems guide the operator through the setup, permit the progress of the analysis to be monitored, and assist in the preparation of reports and storage of results. They can provide assistance to the operator at any time in the form of help screens and can also help provide the expertise required to make decisions regarding operating conditions for a given method and the quality of results. These intelligent interfaces should be freeing analysts to work on problems associated with existing methods and develop new methods.

Figure 8.6 shows a schematic diagram of a fifth-generation instrument.

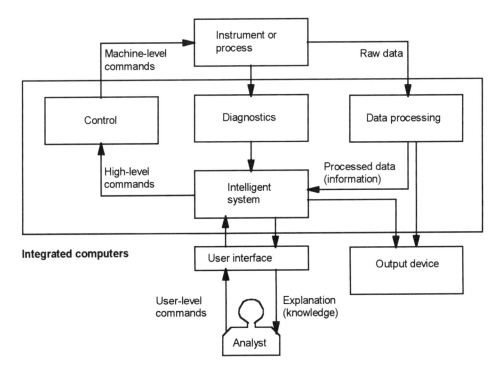

Fig. 8.6   Fifth-generation systems.

8.3 SYSTEMS VALIDATION AND CREDIBILITY

Future generations will incorporate advances in both hardware and software to provide systems that are even more efficient, flexible, and intelligent. They will have new sensor technology to improve specificity, increase sensitivity, and lower detection limits. Some progress has been made in this respect by a number of workers using microwave sample preparation techniques on-line, for example,

However, in relation to sample-preparation techniques there have been few advances, and this remains the major stumbling block and area for development. The experience to design good systems in this respect is difficult to obtain and requires a good understanding of the chemistry involved and an ability to appreciate the precise requirements and the necessary constraints imposed by the sample, its source and the legislation involved.

Haswell and Barclay [3] have described a microwave system coupled to an atomic absorption detection system for the analysis of sludges and soils. A major constraint at the present time is that the preferred operation of these types of systems is for sample matrices to be closely matched. A widely varying sample, which exhibits different heating characteristics, will either show up as an invalid result or the time required to cope with this procedure for all the samples will greatly extend the on-line analyses time scales. As more of these instrumental systems become linked to laboratory information management systems, it will become feasible to interact between the control database and the instrumentation so that each sample is treated in an appropriate manner and the optimum time frame is selected for each sample type. When new samples are analysed, the steps could be monitored so that the required time scales are obtained and then stored for future reference.

The quality of analytical data has been addressed by Taylor *et al.* [4]. Analytical chemists must be able to measure analytes confidently in the matrix of interest. Results must be meaningful and reliable. This involves interaction from the outset in deciding what samples are to be taken and where they should come from. Analytical procedures must be clearly defined. Analysts must take pride in their achievements and should involve themselves in all aspects of the measurement programme, including the way results are reported and communicated to the public at large. Automation has a significant part to play in this respect. There is a great need for reliable automatic systems that give results in which the end-user will have confidence.

Automation has a key role to play in providing adequate and reliable traceability of results. This does not simply mean that every sample must have a bar code to identify it. Other techniques and control systems are often less expensive and more reliable. Ease of operation and control are the major points to be considered here. Simple computer systems with good user friendliness are available and these should be used. Many computer programs do not take the user of the system into account, and their designers should start looking at what motivates the user to operate these correctly and easily. Laboratory managers must be aware of the views of their junior staff, and most decisions on automation must be taken at all levels within a company. A record system must be usable both by the present staff and by future recruits. How often have you heard that after a series of shift changes, the operation of the system has fallen down? The first operator develops discrepancies which get passed onto the next and so on. After a year or so, the system collapses and is no longer reliable.

Accreditation schemes have lately found much favour and it is expected that use of such schemes will increase. Unfortunately, at the moment, these schemes are reduced to mere indications that defined procedures have been followed, and they do not guarantee that reliable results are always achieved. It is to be hoped that such schemes will be enhanced to give full confidence to the laboratory customer.

Analytical chemistry has to embrace all aspects of a particular problem. Analysts are frequently specialists in chromatography, or even more narrowly specialized in gas chromatography, techniques which cover only a part of a problem, i.e. separating the sample into its component parts. There is a danger that a specialized analyst will predefine the solution to the problem and not select the most reliable and accurate measurement procedure for the analysis in hand. To avoid this, staff need to be trained in the wider skills of analysis.

As instrumentation becomes more sophisticated and subsequently detection levels reach lower and lower levels, considerable demands are placed on the suppliers of chemical reagents, standards and even on sample receptacles. Currently, detection levels in the region of 1 in 10^{-12} are common, and it is often difficult or impossible to obtain reagents which do not contribute significantly to the blank levels. In the field of mercury analysis this is particularly important. Here, if the detection levels were to be reduced even by only one more order of magnitude, then it would be almost impossible for any analyst with even a single mercury amalgam filling in his or her teeth to be in direct contact with the sample and/or analytical instrumentation.

Also, almost as a direct result of the introduction of automation and the subsequent improvement in detection capabilities, the analyst can be drawn into a false sense of security. Analysing at such low levels is difficult problem. Collecting a sample which is representative of the matrix under test and which does not become contaminated or lost before the measurement is very difficult. It requires proper attention to detail, and good, thorough training and education. It is vital that these skills and requirements are not lost as the level of automation in laboratories or on site increases. Without the skill to recognize and deal with these problems, the total benefits of automation will become totally lost.

It is clear that the separation of the sample from the process and its movement into the laboratory for analysis is often a problem which involves time and separates the reason for the result from the analysis. As progress in automation is made, quicker results in-house, which give a clear indication of how the process is proceeding, will have a greater value than a rigorous laboratory procedure which delivers a result after an hour or more. Automatic on-line results should be indicating a trend, rather than providing a precise analysis. Monitoring rivers and outflows using instruments on site, for example, would obviously avoid delays, as well as transfer problems. Future efforts to produce reliable sensor systems will ultimately give more control over environmental pollution.

Research in analytical chemistry is clearly an area where automation has a significant role to play. It is important that research data is fully validated and as accurate as possible. While it is not always possible to automate entire processes, the use of automated carousels to feed samples into a reaction system is an obvious area to improve the quality and rate of generation of data. It will also allow the researchers to quickly validate their proposed methodology on real-world samples and optimize the performance characteristics. This naturally requires a very close relationship between the researchers and the ultimate end-users of the analytical product. Given a good return for the investment, I am sure that the initial investment to automate the research activity will be justified and forthcoming.

In Chapter 1, an attempt was made to define laboratory automation or 'automatic chemistry'. Denton [5] has given an unusual definition which outlines the complexity of the subject matter:

'A multifaceted and multidisciplinary problem, the various facets of which are fully interlocking'.

As analysts become more aware of the fact that automation offers a variety of tools, such as robotics, computing, and statistics, to solve analytical problems, then future progress will be far quicker, allowing the installation of more successful and reliable automatic systems.

Process chemists, clinical chemists, and industrial chemists have similar problems and each should be looking outside his own discipline to effect a cross-fertilization of ideas and concepts. As new sensors and techniques are developed and become available, system designers must have the flexibility to adapt them to their own environments and provide reliable analytical solutions.

8.5   CONCLUSIONS

This book has been written with two principal objectives in mind. First, to prompt analysts to think about their problems, and second, to review objectively how to solve them. The author has drawn on his experience to open up the subject to readers. If his examples have a specific application for you, that is a welcome bonus. If they simply force you to review the possibilities of automation, the author will have made a contribution to ensuring a real and definable analytical service providing unbiased, quality data in a realistic turn-around time.

REFERENCES

[1]   National Instruments Corp. (UK) Ltd, 21 Kingfisher Court, Hambridge Road, Newbury, Berks RG14 5ZX, UK.
[2]   Settle, F.A. Jnr., *American Laboratory*, November, 1993, 15.
[3]   Haswell, S.J. and Barclay, D., *Analyst*, 1992, 117, 117.
[4]   Taylor, G.L., Smith, T.R. and Kamla, G.J., *Journal Automatic Chemistry*, 1991, 33, 3.
[5]   Denton, M.B., Personal communication, 1981.

Glossary of manufacturers' addresses

Allied Data Ltd, Hetton Lyons Industrial Estate, Hetton, Houghton-le-Spring, Tyne & Wear, UK.

Alpkem (product reference available from Perstorp Analytical Alpkem, see separate entry).

Anton Paar, Kärntner Strasse 322, Postfach 58, A 8054 Graz, Austria.

ARL (Applied Research Laboratories), (see Fisons Instruments).

Arizona Instrument, 4114 East Wood Street, Phoenix, Arizona 85040-1941, USA.

Baird Corporation, 125 Middlesex Turnpike, Bedford, MA 01730, USA.

Bayerisches Landesamt für Umweltschutz (BLU), Postfach 81 01 29, 81901 München, Germany.

Beckman Instruments Inc. (product reference available from Fisons Instruments, see separate entry).

Bran & Luebbe (formerly Technicon Instruments), Werkstrasse 4, W-22844 Norderstedt, Germany.

CEM Corporation, 3100 Smith Farm Road, Matthews, NC 28105, USA.

Chemlab Scientific Products Ltd, Construction House, Grenfell Avenue, Hornchurch, Essex RM12 4EH, UK.

Cobas Fara (product reference available from Roche Diagnostics, see separate entry).

Coulter Corp., P O Box 169015, MC 195-10, Miami, FL 22116-9015, USA.

Dionex Corp., 1228 Titan Way, Sunnyvale, CA 94088, USA.

Dunlop Aviation Group (product reference available from Dunlop Equipment Division, see separate entry).

Dunlop Equipment Division, Holbrook Lane, Coventry CV6 4AA, UK.

Eastman Kodak Co., 1001 Lee Rd, Rochester, NY 14652-3512, USA.

Filtrona Instruments & Automation Ltd, Rockingham Drive, Lindford Wood East, Milton Keynes MK40 6LY, UK.

Fisons Instruments (incorporating VG Elemental and Applied Research Laboratories), Ion Path Road Three, Winsford, Cheshire CW7 3BX, UK.

Floyd (product reference available from Oi Analytical, see separate entry).

Hewlett Packard, P O Box 533, 2130 AM Hoofddorp, The Netherlands.

Glaxo, Harmire Roard, Barnard Castle, Co. Durham DL12 8DT, UK.

Index Instruments Ltd, Bury Road Industrial Estate, Ramsey, Huntingdon, Cambridgeshire PE17 1NA, UK.

International Federation of Clinical Chemists (IFCC), Centre du Medicament, Universite de Nancy 1, Nancy, France.

International Union of Pure and Applied Chemistry,

Ismatec SA, Feldeggstrasse 6, CH 8152 Glattbrugg, Switzerland.

Jerome (product reference available from Arizona Instrument, see separate entry).

Knapp Logistics Automation GmbH, Gunter-Knapp Strasse 5-7, A-8042 Graz, Austria.

J E Meinhard Associates Inc., 1900-J E Warner Ave, Santa Ana, CA 92705, USA.

Mettler Toledo Inc., 69 Princeton-Hightstown Rd, P O Box 71, Hightstown, NJ 08520 - 0071, USA.

National Bureau of Standards, Gaithersburg, Maryland 20899, USA.

National Instruments Corp. (UK) Ltd, 21 Kingfisher Court, Hambridge Road, Newbury, Berks RG14 5ZX, UK.

Oi Analytical, P O Box 9010, College Station, TX 77842-9010, USA.

Olympus Corporation, 4 Nevada Drive, Lake Success, NY 11042-1179, USA.

Perkin Elmer Corporation, 261 Main Avenue, Norwalk, CT 06859 - 0001, USA.

Perstorp Analytical Alpkem, 9445 SW Ridder Rd, Suite 310, Wilsonville, Oregon 97070, USA.

Portescap (UK) Ltd., Headlands Business Park, Salisbury Rd, Ringwood, Hampshire BH24 3PB, UK.

Procatalyse, 212-216 Ave. Paul Doumer, 92500 Rueil Malmaison, France.

Prolabo, 12 Rue Pelee, BP 369, 75526 Paris - Cedex, France.

P.S. Analytical Ltd, Arthur House, Unit 3 Crayfields Industrial Estate, Main Road, St Paul's Cray, Orpington, Kent BR5 3HP, UK.

Ringsdorff-Werke GmbH, Bad Godesberg, Drachenburgstrasse 1, 5300 Bonn 2, Germany.

Roche Diagnostics, F. Hoffman-La Roche Ltd, CH 4002 Basel, Switzerland.

Skalar Analytical BV, P O Box 3237, 4800 DE Breda, The Netherlands.

Solomat Manufacturing Ltd, 2 St Augustines Parade, Bristol, UK.

SpinOff Technical Systems, 1 The Avenue, Hadleigh, Benfleet, Essex SS7 2DJ, UK.

Technicon Instruments (see Bran & Luebbe).

Thermo Jarrell Ash Corp., 8E Forge Pkwy, MA 02038, USA.

Varian Australia Pty Ltd, ACN 004 559 540, Mulgrave, Australia.

VG Elemental (see Fisons Instruments).

Vickers Ltd, (this company is no longer trading).

Yorkshire Environmental LabServices, Templeborough House, Mill Close, Rotherham S60 1BZ, UK.

Zymark Corporation, Zymark Center, Hopkinton, Massachusetts 01748, USA.

Index

Note: the following abbreviations have been used. GC – gas chromatography; ICP – inductively coupled plasma; MS – mass spectrometry.

absorbance 39, 40, 53–4, 82, 188–90
accreditation 98, 226–7
acid 75, 104, 105, 123, 129, 147
    dispensing system, programmable 129
actuator 104, 185, 186
adsorber 88, 92
aerosol 138, 186 8
air bubble 48, 49, 50, 54, 55, 104
air-segmented analyser 49, 105
alarm 35, 36, 44, 84, 116
alcohol analysis 10, 13, 104, 105
Allen-Bradley programmable logic computer 178
Allied Data Ltd. (U.K.) 196
Alpkem RFA300 55
aluminium 150
    heating block 175, 181
American Chemical Society 23
amino-acid analyser 30
analogue input/output 14, 21, 223
analysis
    steps 6, 8
    time 7, 96, 155, 200
analyst, chemical 9, 223, 226, 227
    role 5–6, 9, 16–17, 60, 68, 85, 221
        specification of requirements 12, 25, 26
analyte 47, 62, 137, 147
Analytical Nomenclature, Commission on 5
analytical
    chemist *see* analyst
    chemistry, conference on 23
    procedure 126, 226
animal 41, 42
    feed analysis 63
Applied Research Laboratories 140
archiving 196
argon 157, 143–4, 201
    sheath 143, 207
    stream 91, 143, 156, 157, 207, 209
arsenic 43, 131, 139, 141, 142, 171
artificial intelligence 165
ASA Aroma System 198, 199
ashing 120, 127, 129–32, 156, 171
atom reservoir 137

atomic
    absorption 89, 141–3, 145, 146, 171, 226
        spectrometry (AAS) 125, 131, 137, 148, 150, 158
        see also under flame
    fluorescence 89, 90, 143, 145, 147, 211, 213
    spectroscopy 95, 137–59
audit 98, 99–100
AutoAnalyzer 8, 9, 14, 48, 49–57, 81, 83, 105, 141
Automatic Dilution 57
Automatic Wet Digestion VAO system 128
automation 5, 6–10, 27–8, 168
autosampler 21, 55, 143, 202, 210, 217, 219

back pressure 92, 117, 196
back-flushing 107, 108, 109
Baird Corporation 153
balance 83, 122, 188
bar code system 35, 176, 177, 198
batch
    analysis 8, 62, 115, 128, 152, 212, 219
    hydride generation 139, 140
Beckman Instruments (USA) 194, 201
Beer's Law 52
Benchmate product 22, 133
beverage analysis 11, 104, 105, 106
Bico model 6R vertical grinder 171, 172
biochemical oxygen demand (BOD) test 95, 96
biochromatography 114–15
biological
    matrix 166, 170, 171
    sample 127, 129, 131, 168, 171, 178–9
Biomek 1000 167, 194
blood analysis 21–2, 28, 29, 30–1, 55, 137
botanical material 170, 178–9
Bran & Luebbe 62, 82
    see also Technicon Instruments; TRAACS
BRANDER control program 84, 85
bromination 208, 209, 214, 215
bubble 48, 49, 50, 54, 55, 104, 110
    see also debubbler
burning zone 71, 72

calibration 25, 41, 55, 89, 92, 132, 171
   curve 54, 56, 143, 204, 209
   drift 146
   range, linear 201
   regulation 83
   standard 9, 54
carousel 131, 227
carrier stream 108, 116, 143, 201, 209
carry-over 49, 50, 55, 56
cavitation 138
CEM Corporation (USA) 129, 130
centrifugal
   analyser, integrated, robotic 40–2
   mixing 36, 37
   separation system 103
centrifugation 102, 178, 179
cereal product analysis 62, 63, 132
change-over 9, 42, 57
chelation 121, 124–5
chemical oxidation demand (COD) 95, 96
chemiluminescence 96, 180–2
chemist *see* analyst
Chemistry, International Union, Pure and Applied 5
Chemistry, Journal of Automatic 5, 27
Chemistry, Royal Society of 23
chemistry, automatic, definition of 228
   see also under analytical; clinical; wet chemistry
ChemLab Scientific Products (CSP) 218, 219
ChemStation 119, 120
ChemStore database 119
chlorinated compound extraction and analysis 183–6
Chromasorb 75, 109
chromatography 20, 107–20
   see also under gas, high pressure liquid, liquid
cigarette analysis 10, 23–4, 67–86
   see also smoke; tobacco
cleaning 55, 126, 172, 173
   see also wash; washing
clean-out 171, 203
Clinical Chemistry, International Federation of 5
clinical
   application 26, 47, 191
   chemistry 25, 166
   market 9, 19, 21, 22, 42, 50
closed-vessel technique 129–32, 170
Cobas Fara II 40–2
cold vapour atomic absorption spectroscopy 89, 145
colorimetry 9, 28, 31, 55, 82, 110–11, 145
colour 96, 195, 197, 198, 200, 218, 219
   density 46, 47, 82
combustion 71, 127, 129, 180, 182
communication 26–7, 170, 221
COMPAQ 386s microcomputer 179, 186
completely continuous-flow analysis (CCFA) 48
computer and computing 5, 15, 19–21, 23, 163–91
   control 43, 55, 80, 120, 207
   power 41, 221–5
   software *see* software
   see also individual manufacturers

concentration 53–4, 61, 162, 203
condensation 158
conductivity 96, 104, 217–19
conference 23, 133
constraint 26, 67, 121
contamination 61, 120, 121, 122, 126, 170
   cross 29, 173, 200
   see also interaction; interference
continuous-flow
   analysis 8, 9, 14, 19, 28–9, 60, 212
   automatic analyser (CAA) 47–62
   process 13, 30, 47, 50, 123, 128, 153
   extraction (CFE) 62, 102, 114, 138–9
   hydride/vapour generation 140, 141
   segmentation 48, 49
   separation system 47, 50
Contract Laboratory Procedures 32
Control of Substances Hazardous to Health Regulation 1989 (COSHH) 97
control system 43, 56, 125, 156, 157, 179, 210–11
controlled-dispersion flow analysis (CDFA) 48, 49
Cool Plasma Asher CPA4 129
cooperation 12, 17, 25, 27, 42, 67, 85, 164
corrosive substance 29, 121
cost 7, 9, 14, 26, 46, 117, 129, 186
Coulter Scientific Incorporated (USA) 40
customer 15, 25, 78, 100, 167, 227

data
   acquisition 20, 203, 225
   collection 50, 77, 84, 210–11, 218, 221, 222
   management 40, 84, 117, 119, 203–4, 222
   presentation 6, 13, 35, 41, 204–5, 206, 210, 223
   processing 6, 9, 20, 67, 68, 76, 113, 225
   quality 170, 226
   retrieval 41, 77, 186
   storage 42, 56, 77, 186
   validation 35, 77–8
   volume 19, 68, 75
database 69, 80, 119, 226
debubbler 50, 53, 55, 110
decomposition 32, 126, 127–32, 170, 171
density 195–200
design 8, 9, 13, 25, 83, 120
   in-house 12, 13, 25, 26
   instrument 17, 22, 54, 196–7
detection 50, 51, 60, 98, 104, 126, 205
   limit 138–9, 141, 150, 156, 227
      improvement 137, 142, 201, 212
diagnostics 118, 224
dialysis 55, 82
digestion 8, 31–3, 120, 123, 125, 170, 171
dilution 28, 29, 61, 153–5, 170, 181, 182
diode array detector 98, 205
Dionex 148
direct-current plasma 141–2, 143
   emission spectroscopy (DCP-OES) 200–4
Discrete Analyser with Continuous Optical Scanning (DACOS) 38–40
discrete analysis 28, 29–36, 38–40
dispenser 44, 179, 189, 190, 191
dispersion 55, 60, 61, 62, 147, 153, 201

Index

display 35, 41, 201, 203, 206, 223, 225
dissolution 127, 182, 204
    tablet 54, 113, 182–3, 204–7
distillation 13, 55, 75, 80, 81, 83, 104–6
documentation 76–7, 80, 84
dose delivery system 187
downtime 12, 89, 182
Dreschel bottle 89
drift 54, 55, 56, 110, 146, 155
    baseline 54, 146, 203, 304
Drinking Water Inspectorate 99
dual electron capture (ECD) detector 184
Dynamic Data Exchange 119

Eastman Kodak 46
Ecole National Superieure des Mines de Paris 164
economic factor 7, 8, 11, 27
education 17, 22–4, 123
effluent 96, 115, 150, 183–6, 208
electrode 9, 40, 41, 44, 104, 124, 217
    conditioning solution 45
electrodeposition 148
electronics 9, 12, 23, 157
electrothermal vaporization (ETV) 148, 150, 155–9
elution 75, 149, 150, 153
entry station 43, 44
Environment, Department of: water report 219
Environmental Protection Agency 32, 42, 173
environmental test 15, 42, 148, 173–5, 227
error 31, 76, 99, 153, 182, 187, 201
    systematic 99, 126, 131
Ethernet™ network 21
European Commission 10
European Union tobacco duty system 69
evaporation-to-dryness module 54, 102, 113
event log 79
experimental design 21, 77, 80, 84
extraction 11, 54, 97, 102, 121, 124–5, 178–9
    continuous 62, 102, 114, 138–9
    liquid-liquid 101–4, 148
    solvent 11, 12, 29, 55, 101–4

failure 8
feedback 9, 169
filtration 12, 73, 75, 114, 214
Filtrona 300 smoking machine 68, 72, 84
Fisons Elemental 140
Fisons Instruments (UK) 144
flame atomic absorption spectrometry (FAAS) 137, 145, 148, 152
flash distillation 13, 81, 104–6
flow
    cell 50, 56, 59, 205
    rate 55, 60, 62, 110, 123, 137, 155, 201
flow-injection
    analysis 14, 48, 57–60, 137, 200, 201
    application 23, 145–53
    approach 61, 140, 143, 146, 201–3
Floyd Incorporated (USA) 129, 130
fluorescence 40, 41, 90, 145, 146, 209
    detector 91, 144, 182, 207
food analysis 13–14, 42, 120–5
fragrances and flavours industry 195–200

Galahad, Sir, system 89–94, 211
gas
    natural 86–94
    oil 11, 12, 107
gas-chromatography (GC) 80–1, 97, 111–20
    GC-MS 98, 183–6
    hybrid system 107–11
gas-liquid
    chromatography 75
    separator 143, 144, 207, 211
GEC Robot Systems RAMP 2000 166
General Medical Sciences-Atomic Energy Commission (GeMSAEC) 36–8, 102
geological sample 171–3
Glaxo Operations (UK) 205, 206
gold trap 89, 91, 92, 145, 210, 213
Good Laboratory Practice (GLP) 56, 98, 99, 119, 120, 198, 204
Government report
    laboratory 56, 98, 99, 119, 120, 198, 204
    smoking 13, 67
    water 8, 97, 98, 99, 219
gravimetric analysis 173–5
Graz University 132, 148
grinding 169, 171–3, 178, 179

Hage-Poiseuille Law 112
half-wash time 50, 51, 52, 53
Hall Effect detector 190
'hands-off' analysis 9, 54, 168, 182, 198, 210
hazardous material 97, 165, 168
Health, Department of 78
    see also Good Laboratory Practice
heating 31, 45, 50, 128, 129–30, 156, 158
    controller, proportional 123
Hewlitt Packard equipment 117, 118, 120, 206
    ChemStation 119, 120
High Pressure Asher 129, 130, 131
high pressure liquid chromatography (HPLC) 14, 54, 57, 58, 97, 98, 205, 206
hydride system 121, 131, 139, 141–2
hydride/vapour generation 139–45, 210
hydrocarbon 97, 107, 108, 188

IBM 176, 181, 182, 196, 206
Index Modular Analysis System 200
inductively coupled plasma (ICP) technique 20, 96, 126, 137–47, 155–9
inert material 29, 121, 122, 126, 200, 215
InfraAnalyzer 400 63, 123
infra-red reflectance technique 63, 132
injection 57, 59, 83, 111, 116, 117, 171
    device 109, 149, 150
InnovaSystems Incorporated 187
inorganic
    decomposition 130, 131
    material 126, 214
inspection 99, 100
Instrument Control Language (ICL) 21
instrument
    automatic, principles of 28–47
    design 17, 20, 22, 54, 196–7
intelligent automation 202, 225
interaction between samples 29, 51, 53
    see also contamination; interference

interface 102, 122, 137–9, 158, 170, 223, 225
   man-machine 19, 43–4, 226
interference 50, 55, 82, 143–4, 150, 152, 212
   element 141
   see also contamination; interaction
Ismatec (UK) ASA Aroma System 198, 199
ISO 6978 potassium permanganate method 93

Jerome 431-X Mercury Analyzer 88, 89
Josco Smartarm 167, 194

Kel-F 23, 121, 122
Knapp Logistic Company (Austria) 124, 148
Konig reaction 81

Laboratory of the Government Chemist (LGC)
    10–13, 27, 50, 104, 120, 166
   Laboratory Automation Group 12, 67
   on tobacco 7–8, 67–100
Laboratory Information Management System
    (LIMS) 95, 170, 225, 226
laboratory 9, 10, 25–6, 27, 57, 99, 191
   management 17, 25, 69, 226
   research 165, 166, 227
   station 133, 174
   unit operation (LUO) 169, 178
   wet chemical chemistry 43, 44
LabVIEW 20
lag pahse 40, 50, 52, 53
large volume on-column injection (LOCI) 183–6
lead 137, 139, 148, 158, 159
   in drinking water 12, 96, 148, 150
legislation 26, 95, 96, 207, 208, 221
linearity 182, 183, 210
liquid sampling 28–47, 117, 119
liquid-chromatography 98, 113–14, 137
   see also high pressure liquid chromatography
liquid-liquid extraction 101–4, 148
literature 27, 57, 62, 104, 141, 165, 171
   manufacturer 17, 25
loading 31, 56, 131, 149, 150

maintenance 8, 29, 30, 35, 118, 179
Malmstadt, Professor H. 5
management 9, 10, 17, 25, 69, 99, 164, 226
manifold 55, 60, 146, 148, 150, 156
   design 52, 147, 149, 208
man-machine interface 19, 43–4, 226
mass
   spectrometer 20, 21, 126, 144
   spectrometry 97, 98, 143, 146, 183–6
   ICP-MS 96, 139, 143, 147, 153, 155–9
Master Laboratory Station 174
Masterflex® peristaltic pump 117
MasterLab 167, 194
matrix
   biological 166, 170, 171
   effect 21–2, 31, 180
   interference 141, 146, 201
Maxx-5 167, 194
Meinhard nebulizer 137
membrane 82, 106, 115, 137–9, 143, 219
mercury 8, 86–94, 139, 144–7, 214, 227
   low level 146, 207–17
   vapour 143, 207

Merlin system 89, 90, 92, 207, 211
metal analysis 33, 95, 96, 131, 200–4
metered aerosol product 186–8
method
   detection limit 96, 175
   development 153, 179
methods management system 169
Mettler equipment 22, 43, 83, 122, 188
Microdigest system 128, 131, 132
Microlab 2000 167, 194
Microsoft software 19, 179
Microwave Acid Digestion Bomb 129
Microwave Digestion System MDS-81D 129
microwave
   digestion 128, 129, 131, 132, 170
   process 96, 102, 128, 130, 131, 171, 226
milk and milk product analysis 63, 113
Minimover-5 167, 194
Mini-S rinse pump 199
Mitsubishi Move Master 167, 194
mixing 28, 47, 122, 124, 175, 191
modular analysis system 22, 23, 118, 200
monitoring 91, 97, 126, 227
monochromator 39, 201
Multichannel 300 analyser 30
multichannel instrument 8, 11, 30
multi-layer chemistry 46–7

National Bureau of Standards (NBS) 125, 158,
   159
National Instruments 20, 222
National Measurement Accreditation Scheme
   (NAMAS) 98, 99
near infra-red reflectance spectroscopy 6, 62, 132
nebulization 139, 140, 142, 156, 201
nebulizer 137–8, 149, 153
nephelometry 40, 219
networking 21, 119
neutralization unit 121, 124, 125
nicotine 23, 67–86
nitrogen 108, 110, 180–2, 209
   robot 180, 181, 182
non-dispersive infra-red carbon monoxide
   specific detector (NDIR) 74, 84
nuclear magnetic resonance spectrometer 20
Nutrients Composition Laboratory, USDA 10,
   13–14
nutrient 95–6, 97

Oak Ridge National Laboratory (USA) 36
oil 153, 180
Olympus AU5000 series 33–6
on-line analysis 15, 55, 63, 86, 93–4, 132, 153–5
   testing 217
open vessel technique 128–9, 132
operator 9–10, 12, 56
   see also under staff; user
Optical Mark Registration (OMR) system 95
optical
   emission spectroscopy (OES) 131, 152, 153,
    165
   rotation 198, 200
   system 21, 31, 39, 44, 63, 177
organic
   material 98, 127, 130, 131, 208

solvent 137, 138
Ortho Pharmaceutical 170
oxidative combustion 180, 182

Paar, Anton, Company (Austria) 129–30
    equipment 31–2, 196, 198
Parr Instrument Company (USA) 129
password protection 79–80, 120, 179
performance 5, 16, 47, 98, 125
perfume 198–200
peristaltic pump 29, 49, 57, 102, 177, 154, 219
    flow rate 110, 155
Perkin Elmer Corporation (USA) 140, 146, 194
    equipment 126, 142, 143, 166
petrochemical industry 118, 120, 153, 180
pH 57, 96, 124, 169, 217–19
pharmaceutical industry 120, 133, 204
phase boundary
    sensor 104, 124, 125
    separator 102
Philips Research Group 12
photographic process 28
photometry 33, 35, 56
photomultiplier 39
Physicians, Royal Society of 67
pigment 198–200
pipette 40–1, 48, 170, 178, 179, 181
piston 49, 57, 72, 125, 196
plant tissue protein extraction 178–9
plasma 137, 143, 156, 157, 201, 203
plasma-emission spectrometry 125
pollution 148, 217, 227
Polytetrafluoroethylene *see* PTFE
Portescap (UK) Ltd. 104
power and event controller 176, 180, 188
precipitation 25, 148
precision 5, 9, 14, 28, 39, 167, 188, 217
preconcentration 114, 145, 148–53, 211
pre-spectrophotometry system 35
pressure, barometric, changes in 117
pre-treatment 29, 41, 111, 113, 114
printer, Silent 700 83
probe 20, 21, 35, 55, 124, 125, 196
    concentric 154, 155
    platinum-iridium 122
Procatalyse CMG 273 88
process
    analyser 54
    control 19, 21, 22
processing
    plant 9, 10
    rate 28–9, 36
productivity 57, 84, 168, 186
programmable logic controller (PLC) 179
Prolabo (France) 129
    equipment 128, 131–2
propelling unit 49
protein 41, 46, 178–9
protocol development 69
P.S. Analytical Ltd 140, 141, 145, 153, 154
    equipment 140, 141, 143, 157, 219
    see also Galahad; Merlin; Touchstone©
PTFE 102, 110, 123, 129, 130, 149, 200
Puma 167, 194
pump 14, 59, 110, 123, 150, 154, 199, 219

peristaltic *see* peristaltic pump
piston 49, 57, 196
pumping 28, 29–30, 49, 153, 154

Quadrupole Spectrometer 20
quality 12
    assessment 96, 98, 99
    control 29, 35, 55, 41, 56, 69, 78, 99
        industrial 9, 175
        perfume, fragrance and pigment 198–200
        robot 166, 168
quantitation routine 20, 21
quinizarin 12, 107
Quinizarium system 62

rack system 35, 40, 177
random-access sampler 55, 56
Rank Xerox RX-530 computer 67
rapid analysis system, on-line 86
reaction
    console 31
    rotor, stepping 31, 40
    time 38, 52, 58, 146
reagent 35, 44, 57, 110, 128, 190, 227
    addition 29, 39, 131, 179
    consumption 33, 35, 57, 204, 215
    dispenser 31, 40, 128, 189, 190, 191
recorder, potentiometric 82, 124
recovery 96, 182, 183
reflectance technique 6, 62, 63, 132
refractive index 124, 195–200
regulatory requirement 15, 120, 221
reliability 30, 33, 167
remote operation 90, 91, 170
reporting 6, 21, 69, 84, 119, 221, 225
reproducibility 15, 30, 57, 116–17, 147–8, 182, 183
requirement specification 12, 16–17, 24–6, 120
result 6, 16, 44, 69, 143, 225
    display 19, 38, 216
    see also under data
    real time 56, 227
Ringsdorff-Werke GmbH (Germany) 157
robot 15, 16, 95, 163–91
    arm 22, 185, 188
Robotic Sample Processor 510 167, 194
robotics 5, 14–16, 17, 132, 133, 163–91
rotor 31, 36, 40
Ruzicka, Professor 14

safety 8, 14, 97, 165, 168, 191
sample
    handling 21, 40, 204
    identification 29, 31, 35, 42, 43, 44
    see also bar code system
    indtroduction 120, 121–2, 143, 144, 147, 155–9
    loop 149, 150, 185, 204
    numbers of 95, 168, 204
    origin 25, 70–1
    preparation 56, 97, 101–33, 225, 226
sampling 5, 31, 51, 55, 56
    procedure design 14, 70
    rate 50, 53, 61, 147, 215
    system 49, 83, 93, 117, 119

tobacco smoke 69–71, 76
sand trap, gold-impregnated 90
SCARA Selectively Compliant Articulated Robot Arm 166
scrubbing unit 109–10
security, software 79, 80, 120, 179
segmented flow analysis 48, 55, 60
Seiko Instruments & Electronics Ltd 167, 194
Seiler and Martin robotic workcell 177
selected ion monitoring (SIM) segment 21
selenium 12, 13, 131, 139, 141, 171
sensitivity 54, 56, 83, 110, 145, 210
sensor 31, 117, 177, 187, 223, 227
    gold film 89
    level 35, 41, 44
    robot 165, 166
separation 98, 101–33, 205
separator 137, 138, 143
sequencer, programmable 129
serum analysis 21, 28, 33, 41
Severn Trent Laboratories (STL) 95
shaft encoder tube 39
shaker 185, 187
Shell Development Company (USA) 10, 14–16, 180
Sherwood Data System 218
shut-down 8, 57
silent-hours working 8–9
siphon 37, 38
sipper station 185
Skalar assembly 48
sludge analysis 97, 226
smoke analysis 23–4, 67–86
    see also cigarette; tar; tobacco
smoking machine 68, 71, 72–5, 84
software 12, 19–21, 120, 158, 179, 182, 222–3
    algorithm 54
    cigarette analysis 76, 79–80
    data-presentation package 6
    see also Microsoft; security; Touchstone©
soil analysis 97
solid
    determination, percentage 169
    reagent 188–91
    sample 158
solid-liquid extraction system 102
solid-phase chemistry 28, 46–7, 97
Solomat Manufacturing Ltd (UK) 217, 218
solvent 104, 137, 138, 158, 177, 182, 199
    chlorinated 97
    extraction 12, 29, 55, 101–4
specific gravity 195–200
specification of analytical requirement 12, 16–17, 24–6, 85, 120
spectrophotometer 75
spectroscopy, applied: conference 23
stability 186
staff 7, 9–10, 11, 69, 76, 188, 226
    see also operator; user
stainless steel 122, 200, 215
standard, industry 55, 93, 119, 125, 158, 159
steady state conditions 50–1, 52, 54
stirrer 104, 124
sugar 9, 53, 105
sulphuric acid 121–3, 146

surging 104
syringe 29, 35, 123
System V controller 176, 185, 186, 188
system
    definition and validation 78
    designer 26–7, 67, 68, 221
systematic error 99, 126, 131
systems approach 19, 22, 26

tablet analysis 54, 113, 182–3, 204–7
tar in cigarette 10, 13, 23, 67–86
Technicon Instruments 24, 62
    equipment 55, 82, 102, 132
    sampler 14, 82, 113
    see also AutoAnalyzer; InfraAnalyzer
Teflon 36, 131, 201, 211, 215–16
temperature 8, 32, 72, 117, 123, 130
testing 33, 126, 217
Texas 960A computer 40
Thames Water 218
Thermo Jarrell Ash atomic absorption system 143
thermostat, air 110
thermostatic bath 92, 205
thin-film chemistry 28, 46–7
throughput 29, 117, 128, 129, 130, 204
titration 23, 44, 45, 46, 175–7
tobacco 7–8, 13, 63, 67–86, 108, 132
    see also cigarette; smoke; tar
total
    dissolved solids (TDS) 173, 174, 175
    solids (TS) 173, 174, 175
    suspended solids (TSS) 173, 174, 175
    systems approach 13, 27, 67–100
Touchstone© software 92, 143, 207, 209, 210
TRAACS 800 55, 56
trace
    element analysis 126–7, 139, 140, 153, 170, 171
        extreme 32
    metal analysis 33, 120–5, 147
    preconcentration system 150–1
traceability 167, 226
TraceCon 152–3
Trace-O-Mat 129, 132
Trade and Industry, Department of 98
training 10, 11, 17, 21, 22–4, 227
transient signal 60, 92, 146, 156
transport 31, 61, 137, 217
    mechanism performance 39, 43, 44, 157, 158
tube 29, 49, 59, 110, 123, 154, 158
    diameter 55, 57, 58, 60, 61, 196
    dispersion in 147, 201
turbidity 40, 96, 218, 219
TurboVap
    Evaporator 173, 175
    Workstation 174
turnaround 168, 175, 188
turntable 122, 125

ultra-violet system 55, 97, 182, 205
unattended operation 9, 54, 168, 182, 198, 210
United Kingdom
    course/training 23, 24
    tobacco industry and duty 69, 83

Water Act 1989 97, 99
United States of America 23
    Department of Agriculture (USDA) 10, 13, 14
Universal Machine Intelligence RTX Robot 166
University College Swansea 23
urea 46, 53
urine 21, 29, 41, 171, 208
user 20, 24, 27, 79–80, 226
    interface 19, 225
    see also operator; staff

validation 170, 182–3, 210, 227
valve 35, 108, 118, 149, 150, 176, 202
    dual configuration 147
    manifold arrangement 214
    rotary 31, 57
    Teflon 211, 215–16
vaporization 156–8
vapour-generator system 13, 89, 140, 145, 207, 211
Varian system 140, 143
vessel material 32, 57, 129–30, 131
VG Elemental mass spectrometer 126, 144
Vickers Multichannel 300 30, 31
viscosity 147, 153, 154, 169, 196, 200
viscous sample 55, 155, 158, 177, 181
volatile material 126, 181
voltametric determination 131

warning 35, 36, 44, 84, 116
wash 35, 55, 60, 105, 124, 125
    time 179, 200
    system 35, 45, 185
washout 36, 57, 61, 202
waste
    actuation 188
    chemical disposal 44
water 8, 83–4, 96, 175, 208, 227
    analyser 218, 219
    drinking 12, 56, 96, 99, 150
    laboratory 47, 95–100, 217
    legislation 97, 98, 99, 219
    seawater 56, 96, 211–13
    waste 33, 56, 96, 97
weighing 43, 44, 83, 169
wet chemical
    chemistry laboratory 43, 44
    process 120, 127, 128–9, 170
World Health Organization (WHO) 96

Yorkshire Water Laboratory Services 8, 208

Zymark Corporation (USA) 22, 24, 133, 164, 194
    equipment 173, 176, 177
    Zymate 166, 167, 171, 172, 181, 188, 194
        Laboratory Automation System 133, 164, 165, 173
        robot 182, 183, 185, 186
        System V controller 176, 185, 186, 188